WISDOM OF THE ANIMALS

WISDOM OF THE ANIMALS

Communication Between Animals and the People Who Love Them

Raphaela Pope and Elizabeth Morrison

Adams Media Corporation
Holbrook, Massachusetts

Published by
Adams Media Corporation
260 Center Street, Holbrook, MA 02343. U.S.A.
www.adamsmedia.com

ISBN: 1-58062-484-7

Printed in Canada.

J I H G F E D C B A

Pope, Raphaela.
Wisdom of the animals / by Raphaela Pope and Elizabeth Morrison.
p. cm.
ISBN 1-58062-484-7
1. Extrasensory perception in animals. 2. Telepathy. 3. Human-animal communication. I. Morrison, Elizabeth, 1944– II. Title.

QL785.3 .P67 2001
591.5—dc21 00-050252

Cover illustration by Janice Kinnealy.

This book is available at quantity discounts for bulk purchases.
For information, call 1-800-872-5627.

Contents

Come Into Animal Presence

Come into animal presence
No man is so guileless as
the serpent. The lonely white
rabbit on the roof is a star
twitching its ears at the rain.
The llama intricately
folding its hind legs to be seated
not disdains but mildly
disregards human approval.
What joy when the insouciant
armadillo glances at us and doesn't
quicken its trotting
across the track into the palm brush.

What is this joy? That no animal
falters, but knows what it must do?
That the snake has no blemish,
that the rabbit inspects his strange surroundings
in white star-silence? The llama
rests in dignity, the armadillo
has some intention to pursue in the palm-forest.
Those who were sacred have remained so,
holiness does not dissolve, it is a presence
of bronze, only the sight that saw it
faltered and turned from it.
An old joy returns in holy presence.

—Denise Levertov

Acknowledgments

Without the kindness, interest, and participation of many people and animals this book could not have been created. We would like to thank Mary Argo, Sally Blanchard, Carole Boyd and Duke, our agent Sheree Bykofsky, Victor Corbett for his poem on *Peter and the Wolf,* the crew of the *Bottom Time II,* Mary Getten for her great work with whales and for carefully correcting the manuscript, Donna Grady and Jazz, Paul Harris, Michelle Haseltine, Jasmine Indra, Donna Kachinskas, Gillian Lange and the Lange Foundation, Jill Malnick and Petruccio, Bay Manning, Bonnie McNamee, Ralph Morrison, Dicksie Sandifer and Samantha, Joan Rogers, Windstrike, and Helga, Sierra, Carla Simmons, Penelope Smith, Teresa Wagner, and Christina Zelaya and Bobbie Kitty.

Raphaela would especially like to thank all her students, clients, teachers and colleagues. You have helped me grow as a communicator and pointed me in many new directions.

Finally, we offer astonished and heart-felt gratitude to all the animals, named and unnamed, who have talked to us and helped us over the years, sharing their stories, wisdom, and insight. Animal communication is a great spiritual blessing. For it we are deeply grateful.

Introduction

Raphaela

If you had asked me what I wanted to be when I grew up, I would not have said an animal communicator. I had no idea such a thing existed. And if I had? Who knows? Certainly I was one of those children who pop out of the womb simply loving animals. It was a purely innate and involuntary passion. I can remember falling in love with an enchanting orange tiger cat when I was just three years old. I dogged that poor cat's every footstep and carried him around as much as he would allow. I even stuffed him into a doll carriage and pushed him around the yard. He must have been an amazingly tolerant cat to put up with me.

Somehow, that little girl fascinated with animals grew up to be a full-time telepathic animal communicator. I get to talk with animals every single day. I often feel that I am living the fulfillment of my childhood dream. It did not happen overnight, of course. It was the result of a gradual process of evolution that took over 20 years.

Early in my life I lived with my great aunt and uncle on a turkey ranch and dairy in the Sacramento Valley of Central California. I loved all the animals there, from the tiny turkey hatchlings in the brooder houses to the lovely, sweet-smelling cows in the dairy. One of my earliest memories is of trying comfort the newly born calves who were separated from their mothers when only a few days old.

Looking back, I know that I did communicate telepathically with those calves. I believe that many, if not all, children communicate with animals quite naturally. We lose the capacity as we grow up, usually because it is not validated by the adults around us. I lost my early connection when my great uncle, I'm sure intending only to help me, told me that I was mistaken and that the little calves were not really crying for their mothers, as I knew they were. I was too dependent on him to doubt what he said, so I surrendered my knowledge and with it, my telepathic ability. It was many years before I was able to get it back.

Later in my childhood, my mother, sister, and I lived in Valencia Gardens, a housing project in San Francisco. Valencia Gardens had a strict no-pets policy. It was a depressing and difficult time for me. Then one day, as I wandered through the Mission District, I came across the city's Society for the Prevention of Cruelty to Animals, on Capp Street. I was ecstatic! Here were rooms full of dogs and cats who were yearning for love and contact, just as I was.

At least once or twice a week during the fourth and fifth grades I walked to the SPCA after school and lingeringly visited each cage. I petted dogs through the bars and mentally chose the ones I would take home if circumstances allowed. I hatched the idea of having a huge home in the country and taking in all the stray dogs and cats, giving them love, warmth, and security.

On my tenth birthday, my stepfather gave me the best present of my entire life: six riding lessons at St. Francis Riding Academy, a stable in Golden Gate Park. St. Francis was a beautiful, old fashioned city stable, with a lofty ceiling, airy box stalls, and an indoor arena. It reminded me of the illustrations in my favorite book, *Black Beauty.* I was in heaven after my first lesson. I walked through the barn, stopping to talk to and admire each horse, once again planning which animals I would take with me to my country home. I visited the stable after school during the week. Just walking in and taking a deep breath of the horsy atmosphere was intoxicating.

The lessons were over all too soon, and I began riding rent-string horses at the cheaper barns on the beach south of San Francisco. Escorted trail rides did not exist in those days. Most of the horses were rank, barely under control, and would try all sorts of rude behavior to avoid taking their mostly inexperienced riders out. I rode runaways and horses that rolled under saddle. I was so green myself that I had no idea how to control them. I was scared, but it didn't do anything to lessen my love for horses. I wanted to be with them more and more.

I did whatever I could to be with animals. If I saw anyone on the street walking a dog, I introduced myself and tried to make friends with the dog. I got to know a neighbor who allowed me to walk her cocker spaniel occasionally. On one red-letter day a photographer with a pony dressed in western regalia came to my street. I was so enchanted that I, with at least half the kids in the neighborhood, followed the photographer for blocks, begging to sit on the pony or at least be allowed to hold his bridle.

When I was eleven, my mother bought a house and we got a cat. Not just any cat, but an elegant, brown, velvet-pawed Siamese cat. I named her Ling, which sounded appropriately exotic to my eleven-year-old ear. Ling played, basked in the sun, and slept on my bed at night. I had never experienced such an enthralling animal friendship. We invented a game where I would chase Ling in a circle down the hall, through the kitchen and living room, and out into the hall again. Then I would turn and run the other way, with Ling in hot pursuit, streaking by me as she gained traction on the living room rug. She played in paper grocery sacks left open on the floor, sneaking out a velvet paw at my unsuspecting ankle. She pulled socks out of the laundry and "buried" them in the couch, and played with straws that she fished out of the kitchen drawer. I adored her. No evening or afternoon was complete without her ecstatic purring and cuddling on my chest. When I left home for college, Ling kept my mother company for many more years.

I graduated from nursing school, married, had a child, and bought a house in Berkeley. We got a dog—and what a dog he was! Petey was half Golden Retriever, half Great Pyrenees, a giant teddy bear of a dog with a dense, soft, cream-colored coat and golden freckles on his nose. Once, when I was dropping my son off at school, a child on the playground saw Petey in the car and asked, "Is that a real polar bear?"

I worked nights as a critical care nurse. During the day I spent many hours with Petey and the various horses I was riding at that time. I trail rode in Redwood Park, Tilden Park, and Anthony Chabot Park in the East Bay regional park network, which covers hundreds of thousands of acres. I loved traversing the tiny one-track trails deep in the redwoods and the larger fire trails on the ridges, where I could see the snow on Mount Diablo during the winter.

It was during this period, when my love of animals was finally able to find expression, that I first heard of telepathic communication. I was absolutely riveted by the idea. Could such a thing be possible? Much as I loved animals, and as close as I felt to them, I still thought at first it might be a fairy tale—so completely had I forgotten the telepathic connection I had as a young child.

One day I read that an animal communicator named Beatrice Lydecker was to give a presentation to the East Bay chapter of the California Dressage Society, of which I was a member. It was announced that she would explain telepathic communication and give a demonstration by communicating with several of the horses in the barn. Unfortunately, I had to work the night of her presentation. Afterwards I quizzed my horse friends about it. They agreed that she had been quite accurate in her descriptions of the personalities of the horses she talked to. I determined to try to find out more about this fascinating possibility.

I tried researching telepathic communication with animals in the library, but could only find a bunch of quasi-scientific stuff from Duke University that was no help at all. A year or two passed. I heard of another animal communicator, a woman named Penelope Smith. Determined this time to not let the connection slip, I found out that Penelope lived in Northern California, only an hour or so away from me. I called her and scheduled a consultation in which I asked her to talk with Moose, one of the horses I was riding.

Over the next few years I called Penelope several more times. Each time the information she gave me from the animals was helpful. I now reached the tentative conclusion that telepathic communication with animals was a reality. My desire grew to learn to do it myself.

During one of the consultations I learned that Penelope would be giving a workshop in Oakland at the home of one of her advanced students, Jeri Ryan. The purpose of the workshop was to teach ordinary people like me to communicate with animals, and in fact with trees, flowers, rocks, and all aspects of nature. It seemed like the perfect opportunity, and I signed up to go.

The first day at Jeri's house we learned techniques to quiet our minds and focus on what being an animal might be like. I imagined myself as a horse, and felt the wind in my mane and forelock as I raced and played in my paddock. I felt how wonderful it was to roll on my back in the sand and cover my coat with cool wet mud to keep the buzzing, stinging flies away. I felt powerful and strong, but also peaceful and content.

Under Penelope's instructions we practiced many other techniques that day. Then we were given a homework assignment. She asked us to go home and calmly, carefully, and objectively sit with an animal, domestic or wild, and observe her, sensing her feelings and thoughts. We were to put aside our human preconceptions and try to get a sense of that animal's way of being.

Naturally I chose to do this exercise with Petey, who was then nine years old. I knew Petey was very wise, and I had learned a great deal from him, but I wanted so much to be able to experience his wisdom more directly. I told my family I was doing an exercise, buried the telephone in pillows, and firmly closed the door. Then I sat in a chair and observed Petey lying on a rug several feet away. I noticed his beautiful luxurious coat, his plumey tail, his golden freckles, and his luminous expressive eyes, which were beaming back at me.

Spontaneously I silently thought-spoke to him, "Petey, if you are hearing me, please show me a sign." He immediately got up, came over and sat in front of me and put his paw in my lap.

I was astounded and thrilled beyond words, but still I had doubts. I decided to test it again, in case it had been a coincidence. Again I silently asked Petey for a sign that he was hearing me. Again he placed his paw in my lap. I could no longer doubt his intention to communicate with me and his confirmation that he heard me.

In my profound excitement I looked at the homework assignment again to see what to do next. It suggested, "Ask your animal friend what his purpose in life is." Obediently I turned to Petey and thought-spoke the question to him.

The answer came immediately into my mind, "It's to get you to do this, and it's taken me nine years!" I could almost see him laughing as he said this. It was an amazing, profound, and revelatory experience.

Did I immediately become a skilled and expert animal communicator as a result? No. While I never again doubted the reality of telepathic animal communication, I still had doubts of my own ability that persisted for many years. I would often negate the messages I received unless they were validated by Penelope or Jeri, who by then had become an animal communicator herself.

Because telepathic communications from animals are received as thoughts in one's own mind, it is easy to assume that the thoughts come from within yourself, not from the animal. It took a lot of practice, along with a steady growth in my ability to use my imagination in a receptive way, before I began to feel confident in myself and my telepathic skill. I was also greatly helped by the confirmation I received from animals—confirmation that came in the form of changed behaviors resulting from our communications.

The process took years. I continued working as a critical care nurse, and I almost never talked about telepathic communication in the hospital lest the other nurses and the physicians with whom I worked think I was a loony tune. Still, the word got around. A nurse named Chris bred Russian Wolfhounds, and I did several consultations with her dogs. She told some of the other nurses, and I found myself getting occasional requests for consultations from people I didn't even know. I also did consultations for a few animal rescue organizations, such as Community Concern for Cats.

Looking back on this period I can see that what I was doing was serving my apprenticeship. I was not charging for consultations. I did it for fun; the payoff for me was hearing how an animal's behavior or relationship with his people improved after the consultation. That was wonderfully satisfying. I felt my relationship with animals, always rich and important to me, becoming even deeper.

To hear the animals better I needed to quiet my mind more easily and consistently. I began to meditate more regularly. I had been a yoga student for many years, but had always concentrated mostly on the physical postures. Now I started to take the breathing exercises and meditations more seriously, and practiced them regularly at home.

All this time the animals were busily pushing me along this same spiritual path. I talked with animals who gave me insight into their spiritual nature and awareness, or even their past life connections with their families. I now see my connection with animals as an initiation, an opening to my own spiritual nature.

When my beloved Petey died at the age of 13, I grieved deeply. To comfort me, Petey appeared to me in dreams; he also amazed me by simultaneously appearing to my very linear-minded husband. I realized that Petey had been not just a wonderful animal companion but also a spiritual guide. Indeed, he continued helping my growth as an animal communicator even after he died. My connection with him was so strong that when he told me he had been reincarnated, I could not doubt it. He was now a free-flying hawk in the skies of Northern California. I even felt myself soaring on the wind when I communicated with Petey. These experiences made me feel, personally and viscerally, the reality of life after death.

The closer I came to animals, the more it pushed me to take care of the environment we all share. When I learned that Petey had been reborn as a hawk, I began to study raptors. I learned of their incredible sensitivity to the chemicals we humans use in agriculture and industry. I had always been an organic gardener, but now I became almost fanatical about eliminating pesticides and chemicals from my house. It provoked a real crisis when our attic was invaded by roof rats who were eating the insulation. I tried over several months to get them to leave, but they were having a ball and reproducing like mad. They discovered how to climb down the chimney stove pipe into the kitchen, and the end came when a rat actually popped out at me from the stove. I finally reluctantly agreed to use poison bait, but only after warning the rat population that we were putting it out. Remarkably, I found only two dead rats before the rest of them vacated.

Over the next several years I continued doing animal consultations and taking workshops with Penelope and other teachers. I was still working as a nurse. I even applied to a Family Nurse Practitioner master's degree program; but when I got accepted I wasn't elated and thrilled. My reaction was "Oh no, do I have to do this?" I realized I was almost ready to give up nursing and become a full-time communicator, but I was worried about whether I would be able to support myself doing it. I meditated long and hard. Finally I received the guidance that there are lots of good nurses in the world. There are not so many animal communicators.

I took the plunge, and I haven't regretted it for a minute. Today, the passion for animals I was born with has finally been allowed to become the focus of my life. What I do isn't always easy. Animals have sadness and pain in their lives, just as we do. But it always feels right. It is what I am supposed to do.

Communicating with animals has changed my life. I believe it will change yours too. I hope when you read about the animals in this book you'll turn to your companion animals and begin communicating with them. In Chapter 8 you'll find some ideas on how to get started. But what it comes down to is really very simple. Just sit quietly with an animal and observe her, the way I did with Petey. Put aside your human preconceptions and open yourself to that animal's being. After that, just follow where the animal leads. I promise that your life will never be the same.

Elizabeth

"What do you do?" I asked the smiling woman who had just stepped out of her car and stood stretching herself in my driveway.

"I'm a telepathic animal communicator," she said, and reached down to scratch the head of my Shi' Tzu, Theo.

Up to that moment there had not been a place in my scheme of things for "telepathic animal communicator." I instantly created one. I had been waiting for such a person all my life—I just hadn't realized it. Now she was here, and I wanted to know everything she could tell me. Did she really communicate with animals? How? What do they have to say? And by the way, which of my animal companions (I had three at that time) would she like to talk with first?

Raphaela Pope had stopped by my home in Eureka, California, with her husband, a college classmate of mine. They planned to spend a few days with us, relaxing and enjoying the area. Possibly it was not quite polite for me to besiege her with questions about her profession while she was on vacation. Luckily for me, animal communication is Raphaela's passion as well as her work. She is a gifted teacher who shares knowledge as naturally as she breathes. We went inside and began to talk, and my introduction to animal communication had begun.

I don't recall ever doubting that Raphaela could really talk to animals. She inspires confidence, and I am inclined to believe what seems to me quite self-evident anyway—that animals are at least as intelligent as we are. I always saw truth in the saying that dogs could talk if they wanted to; but they know if they did, they would have to get jobs. It doesn't take a rocket scientist to see which way the free food is flowing.

But Raphaela's animal communication was obviously something more. Lots of people talk to their dogs, but not many stop to listen when their dogs reply. It took me a while to realize that Raphaela's communications were not only very detailed and specific, they also weren't bound by the normal constraints of physical proximity. In other words, the animal didn't have to be in her presence for her to talk with them. After all, if you're using telepathy—speaking mind to mind—why should it matter whether you're in the room or in the next state? In fact it doesn't; most of Raphaela's animal consultations take place while she is on the phone with the animal's person. The animal can be anywhere.

If I'd had any doubts about Raphaela's ability, they were laid to rest decisively when, shortly after her first visit to Eureka, she located a lost cat for a frantic woman who had read about her in the local paper. You can read the story of Christina and Bobbie Kitty in Chapter 1. If it doesn't convince you that animal communication is very real, you may be reading the wrong book.

Less than a year after this first meeting, I sponsored and took Raphaela's first Animal Communication workshop, which was held in my barn. Over that weekend I took a few tentative steps towards talking to the animals myself. Then, as our friendship grew, Raphaela allowed me to assist her and naturalist Mary Getten (who is also an animal communicator) write up a series of interviews they had conducted with a wild orca of the San Juan Islands.

When we worked on the orca interviews, Raphaela came to realize that she could either practice and teach animal communication, or she could write about it. She could not do both with justice; there were simply not enough hours in the day. Yet the animal wisdom she was gathering in her daily conversations with dogs, cats, parrots, wolves, and whales needed to be written down and shared. I offered to work with her and write the animals' stories, and was overjoyed when she accepted.

Here's how I see our relationship. I am Dr. Watson to her Sherlock Holmes. Nothing fascinates me more than hearing Raphaela describe her conversations with animals and their often puzzled or bemused owners. They are like mysteries that her skill as a telepathic communicator, and her sympathy for all beings, enables her to comprehend, if not always to solve.

I mean it as a great compliment when I say that Raphaela's time spent quietly communing with animals has given her something of the quality of an animal herself. She has that ability to be entirely present, the kind of joyful attentiveness that makes being with animals such a calming and centering experience. The word that sums up this quality best, for me, is *listening.* The poet Rilke wrote, in the first *Sonnet to Orpheus,*

> *Animals created by silence came forward from the clear*
> *and relaxed forest where their lairs were,*
> *and it turned out the reason they were so full of silence*
> *was not cunning, and not terror,*
>
> *it was listening.*

The ability to listen is critical to animal communication, and those who practice it must cultivate it assiduously. We humans tend to be more comfortable giving out than taking in. We like the sound of our own voice, but if we can learn to listen we will find a whole new world opening up for us in our relations with animals, and indeed with the whole of nature.

The process can begin very simply, in the give and take of life with our own companion animals. Even in the best-ordered household, there are always issues that arise when different species live together and must accommodate one another. Is it right to leave Rex alone while you're out all day at work? Why does Tiger insist on using the chair instead of his scratching post? Is it all right for the hamster to play with the cat? Where should I board the horse? What do I do with the pig? These are the kinds of questions Raphaela deals with all day long—her bread and butter. By putting them before the animal in question she gets answers that are practical and realistic, if quite surprising at times.

Birds are in a category all their own. We share our homes with birds, but they are not domestic animals—they are wild animals with, at most, two or three generations of domesticity behind them, and millions of years of wild heritage. Can a wild bird ever be happy as a companion animal? Is it wrong to take a bird out of its environment, even, or especially, if we have made the environment uninhabitable? What do the birds have to say about it?

Those wild animals who are still surviving in their own environments raise other questions. What do the whales whose breeding grounds we visit think of our intrusions? Is it right for us to be there? Have they forgiven us for hunting them? Is there any justification for hunting? Is subsistence hunting by traditional cultures more acceptable? Is there really a sacred bond between hunters and their prey?

Finally, there are questions suggested by reintroduction projects, in which animals such as the Mexican wolf, who are close to extinction in the wild but who survive in zoos or breeding programs, are being reintroduced into their former environments. These animals are on the very front lines of the human-animal challenge. I almost said "war," because countless species have paid the ultimate price for our domination of the planet. Can we learn to live once more with wolves? Can we really live without them?

These questions are not new. What's different is that an animal communicator like Raphaela can place them directly in front of the animals themselves. I can hardly describe the relief this brings to every human-animal problem. As long as we're locked inside our own brains, and can only consult members of our own species, any solutions we come up with are necessarily one-sided. Let us hear from the animals. We need to hear their wisdom, lest it disappear forever.

A world where the animals speak is the kind of world I want to live in. It's infinitely deeper, fuller, and more amazing than one where we humans only talk with each other. If you can talk with animals, suddenly there are companions everywhere, ready for friendly or challenging interaction. It may turn out that your dog likes chamber music (mine does) or that your cat cares about your sense of social responsibility. The discoveries don't stop at your garden gate, either. You can go out everywhere and find animals to talk to—hidden, often marginalized, or in a few rare places still magnificently lords of their domain. Wherever they are, they are ready to communicate, and unlike most of us, they will have time to talk.

This presupposes that telepathic animal communication is a gift we all share. Raphaela insists that's so. She says it's an innate ability of all humans, one that most of us stopped using when we were little children and have now thoroughly forgotten. I'm inclined to think that's right, but I also think that some people have more of a gift for it than others.

I happen to be an amateur cellist. I do think anyone who wants to can play the cello. I also think Yo-Yo Ma does it better than other people. We can't all be Yo-Yo, but we can all enjoy music. Susan Chernak McElroy, in her book *Animals as Guides for the Soul*, writes,

> *Developing effective and* consistent *telepathic communication with animals takes a tremendous amount of effort and commitment for most people. . . . Most people don't have the time or the commitment for this sort of work. It takes total focus and total dedication. You have to want it that much.*

After working with Raphaela for several years, I see McElroy's point but I'm not sure she's right. Learning animal communication does take some focus, but the ability can grow quite naturally, just by doing it, in the same way that we learn to talk, at our mother's knee. If you feel drawn to try communicating with animals, I think you should just go for it. Use Raphacla's suggestions in Chapter 8 to get started. Take a workshop or seminar if you can. Regardless of how far you go, it will be rewarding. Evcn a little listening is good for the soul.

There is no doubt that animal communication is, at heart, a spiritual activity. When you begin talking with animals there is a sense of returning to a sacred space you were missing, without knowing quite what it was. This is the sense of "animal presence" expressed so beautifully in the Denise Levertov poem at the front of this book. An old joy returns. The spirituality of animals is constantly before you, but you won't find them pontificating or acting especially "holy." Their spirituality is natural and intrinsic. Sermons do not come from cats. And you nevcr find them piously abstaining from what their animal nature would have them do.

Animals are rather like Zen monks, of whom it is said, "Before enlightenment, chop wood, carry water. After enlightenment, chop wood, carry water." Animals are spiritual beings, and they are also totally present as dogs or cats or wolves or whatever they happen to be. In this book, for example, you'll find cats who act as spiritual guides to their people. They're also *cats*, and they do what cats do. Before enlightenment, catch mouse, flip in air. After enlightenment, catch mouse, flip in air.

Animal communication allows us to project ourselves sympathetically into the souls of other beings. This is what we have done in the stories in this book. Some of the stories are told from the point of view of the animal, as if the animal were actually speaking. In fact, it was. We have been careful not to put words, as it were, into the animal's mouth. Each of the stories where the animal appears to be speaking is built from expressions or feelings that the animal specifically conveyed to Raphaela via telepathic communication.

The longer stories that are in the animal's own voice, such as the story of Bobbie Kitty in Chapter 1, or Theo's story in Chapter 2, come from animals we know well and with whom Raphaela has had many conversations. This gave us the confidence that we could convey the animal's actual personality and way of speaking. In the shorter stories, which come from animals we knew less well, or with whom Raphaela had only a few conversations, we have still stuck closely to what the animal said, and been careful not to embroider on it. We'd like you to know that you aren't hearing from us, you are hearing directly from the animal.

When possible we have used the animal's actual words. Yes, some animals do communicate telepathically in actual English words, although sometimes the way they use words has an unusual, almost foreign flavor. When Marvin, a horse, told

Raphaela, "I have gut sounds," it was striking and funny. Marvin was repeating verbatim words that he has often heard from his veterinarian. It was equally striking when another horse, Duke, talked to Raphaela about the "phantom mare"—a device he was asked to use for breeding purposes and which he had trouble, at first, comprehending. When you hear a horse talking about gut sounds, it's a bit like hearing someone from Mars saying that he has been watching television. You just think, "Really? You have?"

More often, the animals do not use words, but communicate in feelings, pictures and sensory impressions of all sorts. Then it has been up to Raphaela to find words for what the animal was "saying." In such cases we have been careful to let you know that the words are Raphaela's words, not the words of the animal. We will say something like, "the animal seemed to say," or "I got that she said . . . " The words are as close to what the animal was expressing as Raphaela could make them.

When we did not feel we knew the animal well enough to tell his story from his own viewpoint, we did not do so. You will find throughout the book places were Raphaela and I write simply as ourselves. We have not followed the common practice of first and second authors, in which the second author (Elizabeth) disappears behind the persona of the first author (Raphaela.) Both of us are present in this book, and when I, Elizabeth, am the one telling the story, I write from my own perspective, like Watson in the Sherlock Holmes stories. Sir Arthur Conan Doyle may have felt it would be easier for the reader to identify with Watson than with the brilliant, inscrutable Holmes. In the same way, we felt that it might help you understand animal communication if you had the viewpoint of an uninitiated, unskeptical but questioning observer like me.

There are also sections in the book that are taken directly from Raphaela's journals. Chapter 4, on Whales, and Chapter 5, on Dolphins, are set on two different expeditions in which Raphaela communicated telepathically with these animals. So that you can easily tell who's speaking where, each section is identified with the name of the "author"—animal or human, as the case may be.

As we progressed through the book, I actually became more involved in some of Raphaela's interviews with animals. The most memorable instances for me were when I participated in Raphaela's conversation with Yabis, the gray whale who was hunted by the Makah tribe in the spring of 1999; when I "listened in" on Raphaela's interview with the band of California condors who are terrorizing the town of Pine Mountain Club, California; and when I was part of the interview of a Mexican wolf known as #131. These were great occasions for me, and have given me a taste for animal communication. I need to study so I can do more. It's amazing.

Regardless of who is telling the particular story, each one is dedicated to the animal whose story it is. Each one comes from the animal's heart, and each one has its own portion of animal wisdom to give. This wisdom is there for all of us, freely offered by the animals whose world we share. All we have to do is listen.

Chapter One

The Cat's Meow

Bobbie Kitty: My Social Responsibilities

Bobbie Kitty finally had to admit it to herself: she was lost.

This wasn't easy for someone as proud as Bobbie Kitty. Yet the fact was incontrovertible. She had tried three times last night to find the little house she shared with her person, Christina, and each time she had failed.

Three times she had carefully circled the quiet streets of the neighborhood, checking out each back yard and listening attentively for Christina's voice. When she had failed to find her house on the third try she had returned, feeling deflated, to the spot beneath a construction ramp that had been her home for the last three days. There, she settled down to wait for Raphaela to contact her.

She knew Raphaela would be in touch, because she had been talking with her every day since the day she had decided to leave Christina. It had been a relief to have Raphaela available to act as an intermediary between her and her person. Bobbie Kitty had been trying for months to tell Christina where she was going wrong, but Christina, for some reason, couldn't seem to hear Bobbie's clear telepathic messages. Humans could be terribly dense, that Bobbie knew.

While she waited for Raphaela's call, Bobbie decided she may as well give herself a bath; her night of searching had left her coat speckled with dry redwood leaves. Here in this part of northern California, the streets ran through forests of second- and third-growth redwoods, and the stumps of long ago cut old-growth trees.

"I *am* an exceptionally beautiful cat," she thought to herself, partly to subdue her growing anxiety. She carefully groomed her handsome russet and chestnut-striped coat and washed her white face and mittens. Even her tail, which had been partially lost in a forgotten accident, was part of her distinctive charm. She finished her bath and was just stretching out on her bed of damp leaves to take a nap when she heard Raphaela's voice in her mind.

"Bobbie," Raphaela called. "Are you there?"

"Finally!" Bobbie replied to Raphaela, a touch sharply. "You shouldn't have kept me waiting." Always keep humans off balance, was Bobbie's motto; it doesn't do for them to suppose they have any kind of advantage, especially when you need them, as she did now.

"I'm sorry, Bobbie Kitty," Raphaela said humbly. "Christina just telephoned. She said you didn't come home last night. And you promised you would. She's terribly upset."

"So she should be," said the cat. She thought quickly. She disliked admitting she was lost, after she'd been so positive she could make it back to her home any time she chose. What should she do? Launch a new series of demands? Invent some reason why she hadn't gone home last night? It would just postpone the inevitable.

"Is there a problem?" said Raphaela. "You did say you weren't lost."

"I'm not!" She thought back over her last conversation with Raphaela, and this led her to review Christina's offenses once more in her mind.

Christina was not a bad person, really. Like Bobbie herself, she was a beauty, and when Bobbie sat on her lap she felt they made a handsome couple. They were both graceful and strong, and although Bobbie had by now attained a well-seasoned maturity, she liked to think that both she and Christina were in the full flush of youth and beauty.

Granted, Christina's body was, like all humans, flawed by its lack of fur, but Bobbie supposed she couldn't help that. She did possess a magnificent head of thick chestnut hair that was almost as attractive as Bobbie's own.

In addition to her good looks, Christina had some personal qualities Bobbie appreciated. She particularly admired the hauteur with which Christina treated others of her kind. When young men from the university where Christina was a student came to visit, she never paid them that kind of fawning attention that can make dogs and humans rather pathetic. She seemed to know instinctively the cat's wisdom that to hold oneself in reserve is always to attract.

Christina resembled a cat in other ways, too. She was nocturnal by nature, and liked to sit up late at night, reading or listening to music. Then Bobbie would cuddle on her lap and they'd commune quietly together. Other times, Christina would go out late—long after dark. Then she'd return in the early hours of the morning looking exhilarated. "I've been out dancing," she told Bobbie, but Bobbie didn't quite understand, or, in this case, quite approve.

Bobbie stretched once more, trying to get comfortable in what was, after all, only a rough, makeshift bed. She thought wistfully of her pillow at home, but then steeled herself as the thought of Christina's wayward ways returned. It was one thing to treat young men carelessly, and another to behave the same way towards a cat!

The truth was, Christina had become more and more unpredictable in her habits since they'd moved out of the house where they'd grown up in order to attend University. She and Christina had been companions since Bobbie was a kitten, so when Christina went away she naturally took Bobbie with her. Bobbie hadn't especially enjoyed leaving Christina's parents' comfortable home, but at first it had seemed like it might work out. They found a little house not far from the university campus, and Bobbie thought she could be happy there.

She guessed it was Christina's new freedom that led her to depart from their comfortable rapport. For months now Christina had been falling more and more deeply into what Bobbie thought of as an incorrect lifestyle. It began with those late night evenings when she'd meet other students at one of the clubs for an evenings of dancing. Then she began having friends over to the house late at night—boisterous, noisy friends. Bobbie didn't like that at all.

Then Christina brought in something called a "roommate." "She'll help pay the rent, and Tiger will be company for you," she told Bobbie. To her horror, Bobbie realized that Tiger was another cat! She hissed roundly at Tiger before racing upstairs to Christina's bedroom. Things were deteriorating badly.

Roommates were bad enough, but there was even worse to come. Christina began staying out all night. She would leave in the early evening and Bobbie would not see her again until the next morning, or even several days later. This she found quite intolerable. Bobbie had, of course, no problem with staying out all night herself. Christina always called her inside at dusk, but Bobbie exercised her inalienable cat right to do exactly as she pleased. But for Christina to stay out all night was quite another story. It meant she was losing sight of her main priority.

That priority, of course, was Bobbie herself. She expected her dinner promptly at six, and her breakfast no later than nine in the morning. She knew how to mouse, in theory at least, but a mouse or vole here and there could not substitute for a well-filled food bowl. When she'd hear Christina call blithely over her shoulder to the roommate as she breezed out the door, "I'm going to be over at Sarah's for a couple of days. Take care of

Bobbie, won't you?" Bobbie's heart would fill with foreboding. She never knew when she'd get her next meal; the roommate was horribly unreliable.

Bobbie tried—how she tried—to let Christina know how she was feeling. She sent her telepathic messages over and over: "Get rid of the roommate!" "Tiger has got to go!" "You've forgotten my dinner—again!" Christina never seemed to notice them. And to think that Bobbie had imagined they understood each other so well!

At last Bobbie came to a painful decision: she would have to leave Christina. She really had no choice. If Christina realized how badly she'd been treating Bobbie, Bobbie would come back. If she didn't, Bobbie would just have to . . . well, she didn't quite know what she'd do, but she'd think of something.

Once she'd decided to do it, she found that leaving itself was quite easy. Unlike Christina, she didn't have to pack her bags, cancel her magazine subscriptions or notify the utility company. She just had to walk out the door. One day, while Christina was out on one of her infamous walkabouts, she did just that.

It was the next day before Christina realized her cat was gone. She had returned to the house after a late night, and promptly gone to sleep. When she woke up it was mid-afternoon. Bobbie was usually asleep on the pillow beside her at this comfortable hour, so Christina reached out lazily to stroke her soft fur. But her hand encountered nothing. Surprised, Christina got out of bed and went to look for her cat. "That's strange," she said. "I wonder where Bobbie's got to."

When Bobbie didn't appear at six for dinner, Christina began to worry seriously. But she assured herself that Bobbie would certainly be home some time in the evening. By the next morning, Bobbie had still not appeared, and Christina realized that something was badly wrong.

By that evening she was frantic. This was much longer than Bobbie had ever stayed out before. Christina surprised her roommate by bursting into tears and crying, "I don't know what I'll do without my cat!" The roommate had imagined, and not without cause, that Christina barely knew she had a cat.

It was quite true: Christina was desolated. Yet she still had no idea of the reality of the situation. The idea that there might be some connection between her own actions and Bobbie's disappearance did not cross her mind. She assumed that Bobbie was lost or even hurt. She spent the evening walking up and down the streets of the neighborhood, calling "Bobbie! Bobbie Kitty!" over and over.

The next morning Christina resolved to begin searching methodically. First she called the local paper, *The Arcata Eye*, and asked to place a "lost kitty" ad. When she'd finished describing Bobbie Kitty, the woman on the other end of the line asked, "Have you tried Raphaela Pope?"

"Who?" Christina said. "Tried her for what?"

"For animal communication. Wait, let me just find the article." A moment later she was reading out a piece that had appeared in the *Eye* the week before. It seemed that Raphaela had sent the paper an announcement of her Animal Communication Workshop, to be held in nearby Eureka. The *Eye* had run an article on Raphaela with the headline *"Channel Your Chicken's Past Lives, Commune with Your Hamster."* Despite the tongue-in-cheek tone the article was quite sympathetic, describing Raphaela's work in a straightforward, we-make-no-judgments-but-stranger-things-have-happened tone.

"Let me have her number," Christina cried eagerly. Afterwards she told Raphaela that she hadn't experienced a single moment of doubt. Most people, on first hearing about telepathic animal communication, pass through at least a brief "Is that so?" stage. Christina, in her desperation to find Bobbie, managed to skip that completely.

She quickly dialed Berkeley, and a few minutes later she had Raphaela on the line. Raphaela, for her part, heard Christina's tale with a mixture of interest and apprehension. She liked the sound of Christina's voice, and the story of the beautiful lost cat intrigued her. On the other hand, she was still fairly new as a professional communicator, and this would be her first lost-animal case. She'd often heard that these are the hardest cases of all to solve. Many communicators simply refuse to touch them.

She sat for a moment considering the reason, stating it to herself quite frankly. In a lost animal consultation, there's an implacable bottom line: you either find the lost animal, or you don't.

If a client wants to know why her dog is digging in the garden, you have lots of opportunities to work with the person and the dog. You talk to the dog, and he says he digs because he's bored, and can he please have more company during the day. You tell the dog's person this, and he says he'll try to get home from work a little earlier. A week later, if the dog is still digging, you talk to him again and then suggest to his person that he leave the radio turned on. And so it goes.

But with a lost animal, you don't have this kind of leeway. If the animal turns up, you're a hero (and are promptly besieged by a hundred frantic lost-animal clients.) If it doesn't—and there could be a thousand reasons why it might not—it's your fault, animal communication is a hoax, and you are left with only a sad story to think about at night.

Raphaela thought this over and then, with Sherlock Holmesian confidence, decided to take it on anyway. "When did you last see Bobbie?" she asked Christina, and began writing down Bobbie's particulars: what she looked like, how old she was, where she had last been seen.

Then she put down the phone, sat quietly and centered herself. She pictured the flame-colored cat with the half-missing tail just as Christina had described her. Mentally she called out, "Bobbie Kitty, are you there?"

She wasn't surprised when Bobbie answered almost immediately, but she was surprised at the bravado in her tone. If she'd expected a frightened, lost animal, she certainly didn't get one. "Who are you?" the cat asked at once. "If you're calling for *her*, you can tell her I'm not going back. Not unless she makes a few changes!"

"You aren't lost? You're all right?" Raphaela asked, puzzled.

"Certainly I'm not lost. I know exactly where I am, and I'm perfectly fine."

"Hold on a minute, Bobbie," Raphaela said, and picked up the phone again. "I have her," she told Christina. She heard an enormous sigh of relief and a murmured, "Thank God."

"She says she's fine. But Christina, she says she needs you to make some changes before she'll come back. What is she talking about? You haven't been mistreating Bobbie, have you? What have you done?"

"Nothing! I haven't done anything!" Christina cried, sounding aggrieved. "Are you telling me she's gone away on purpose? She can't do that! I've had her since she was a kitten. I'm always so good to her. Unless . . . " Realization seemed to be dawning. "Oh my gosh," she said. "I guess I have been out a lot lately. . . . Maybe I haven't been taking good care of her, but she always seems like she's okay. This is terrible! Raphaela, tell her I'm sorry! If only she'll come back!"

Raphaela went back to Bobbie, expecting the cat to soften at once when she told her how contrite Christina was. She should have known better—this is not the way of the cat. Bobbie knew she had the advantage, at least for the moment, and she pressed it hard.

"Tell her she has to promise *never* to stay out all night again."

"Tell her I won't!" Christina said.

"Tell her I need my dinner on time. And my breakfast. And treats every day," Bobbie added just in case.

"I promise! I'll do anything! Just come home!"

Then, after a pause, Christina added, "You know, this might sound strange but I have had a funny feeling that Bobbie has been trying to tell me something about all this. Would you tell her that, too?"

Before Raphaela had a chance to do this, she heard Bobbie rap out, "Now tell her Tiger and the roommate have to go."

When this was conveyed to her, Christina said, "Raphaela, I can't do that. I need to have a roommate or I can't afford the house. Jamie's a really good one. I know Bobbie hates Tiger, though. Ask her if it would work if Tiger stays in Jamie's room all the time."

Raphaela passed this on to Bobbie, and after an offended silence, the cat said, "Oh, all right, I suppose. Okay. I'll come home. Tonight. Tell her to wait for me—I'm not far away. Have dinner ready."

Bobbie carefully hid her profound relief from Raphaela. Not only hadn't she eaten in three days, she had a tremendous desire to feel Christina's hand stroking her fur once more. But she would be untrue to her cat nature if she let that be known.

Then had come her night of hopeless searching. Now, as she spoke once more to Raphaela, Bobbie felt the advantage had shifted away from her. What *was* she going to do now?

Raphaela was wondering the same thing. She had picked up the phone this morning to hear Christina's frantic voice. "She didn't come! I waited all night for her. Raphaela, you need to talk to her again. Something must have gone wrong!"

At that moment, it must be admitted, Raphaela felt a pang of doubt about her own ability. The truth is that every animal communicator who is honest with herself has moments of doubts at times. The very nature of the process makes it inevitable. Animal communication is so subtle and delicate that in the face of implacable, objective reality it sometimes seems almost to evaporate into thin air.

If Bobbie Kitty hadn't shown up, maybe that meant Raphaela hadn't really been speaking with her at all. Maybe she had tuned in to the wrong cat, or maybe it had all been just a figment of her imagination. In fact, if Bobbie Kitty didn't eventually appear, in the unarguable flesh, there was no way for Christina or Raphaela to be sure that their contact had happened at all.

Then, as suddenly as they had come, Raphaela's doubts vanished. She thought of Bobbie—so quick, so sharp, so clear in her demands—and knew that she could *never* have invented that cat! Once more, she put out the telepathic call, and once more Bobbie responded.

Now, though, as Raphaela listened to Bobbie declaring once again that she wasn't lost, she sensed something in her that she hadn't heard before. There was definitely a hint of fear beneath the bravado.

"Well, if you're not lost," she pressed Bobbie, "why didn't you show up last night?"

"Well . . . " The cat's voice trailed off, but then Raphaela caught a quick mind-picture: a little chestnut cat wandering about in the night, disconsolately searching. So that was it! Bobbie really *was* lost! Why hadn't she said so?

"You couldn't find your home, could you, Bobbie?" she asked her gently.

"I could have! It wasn't there!"

Raphaela suppressed a laugh at this quintessentially cat-like statement: if she hadn't found the house, it was probably because the house had *moved.* She knew Bobbie would hate being laughed at, and she needed Bobbie's full cooperation. Her admission had given the situation a different twist. It was no longer a question of whether she was communicating correctly with Bobbie. Now what mattered was whether she could find a way to bring her home.

Quickly she reviewed what she knew about locating lost animals. Cats do not, as a rule, know the names of streets. They tend to slip around behind the streets, and often prefer back pathways, with their threat of dogs or raccoons, to streets, with their much greater threat of cars. No, you can't ask an animal for its street address. What you can do, though, is ask the cat give you as clear a picture as she can of where, exactly she is. You then describe it to the people, and hope they can use it to locate the cat.

The problem, of course, is that the cat sees from her own point of view: low to the ground, and with a much different focus than ours. Raphaela thought back to the time her own Macaw, Dax, had flown out of the house and gotten lost. She herself had been too distraught to contact Dax telepathically, but another communicator had reached him. He had shown them a picture of a huge brick building, with him crouching near the foundation. When Dax was found, the "huge building" turned out to be Raphaela's own chimney. Like this, the communicator has to somehow decipher what it is the cat is showing her, and then try to translate it into human terms.

Gathering her wits, Raphaela said, "Bobbie, listen to me. I'm going to ask Christina to come and pick you up, but I need you to show me where you are."

"Tell her to hurry. I'm hungry."

"Right, right. Just show me where you are."

Raphaela reached out with her mind towards the wispy, elusive threads Bobbie was sending her. She concentrated hard. There, it was coming in more clearly now. It didn't seem like a very nice place. Damp leaves on the ground. A sort of slanting roof; the sound of hammering; she decided she must be near a construction site of some kind. She looked out from under Bobbie's board roof. There was a strange, red-and-green thing with a wheel spinning in the air—she couldn't figure out what that might be. But that tall pole with the swollen top, backlit by the morning sun, looked strangely familiar. Ah! That was it—a parking meter!

Picking up the phone again, she described the scene to Christina as best she could. Christina was puzzled. The pieces didn't fit. "Bobbie said she was close to home, right?" she said. "I can't think of any parking meters near me. There are some downtown but that's miles away. I hope she hasn't gone that far."

"Think," Raphaela urged. "Bobbie can't get home by herself. If you want her back you have to find her. Are there any new buildings going up near you?"

"Only at the University. Wait a second, that might be it. There *are* parking meters on the streets up there, and it's not far away. I'm going to go out and drive around and look for her. Raphaela, pray for us. I'm off."

Raphaela waited by the phone, following the search as best she could in her mind's eye. She pictured Bobbie, crouching in her makeshift hiding place, hungry, damp, uncomfortable. She saw Christina, upset and worried, driving around on what might be a hopeless search. So this was a lost-animal case, she thought. Like most things, when you got close to one, it turned out to be more complex than you'd imagined.

From one angle, it was about a magnificent being called Bobbie, a cat of great power, more than equal to any human in the depth of her emotions, strong in her determination to have her needs met. It was this cat who had managed, against deep odds, to get through to her person and let her know exactly what she needed from her. She had taken a tremendous, and perhaps irreversible, risk to do it, too.

From another angle, it was the simple story of a lost cat! It was about a small, vulnerable animal, with only a few survival skills, at the mercy of a cold, uncaring world. The cat needed Christina—but not more than Christina needed her cat. For Christina was, finally, just as unfathomable as Bobbie was. She was a beautiful young woman, full of potential, but caught up in the selfishness and heedlessness of youth. She was also a natural animal communicator, Raphaela felt sure; and if she and Bobbie could manage to find each other, she resolved to make sure the two of them learned how to communicate telepathically themselves, so this kind of story would not happen again.

Inwardly, as Christina had asked, she prayed the two would find each other. She wanted so much for Bobbie's brave action to become a hinge on which Christina turned towards becoming the person she was meant to be. Let it not, she prayed, be a tragedy neither she nor Christina would ever forget.

Even as she prayed, Christina was creeping along the streets behind the University, scanning for construction sites on any street with parking meters. Then she saw it—a ramp laid roughly across the sidewalk, exactly the way Bobbie had described it. There was a red and green wheelbarrow upside down next to it, its wheel spinning. Christina drove slowly up to the ramp, came to a stop, and opened her car door. A little head peeked out from under the ramp. It was Bobbie! Even now, Christina saw, she looked absolutely, pristinely, beautiful. With a cry of pure love, Bobbie raced out of her hiding place, leaped into the car, and then, at last, headed straight into Christina's arms.

Elizabeth: Meeting the Divine Cat

I believe the ancient Egyptians were absolutely right when they worshipped Bastet, the cat goddess. Cats do have a divine nature and are quite worthy of all the reverence we can give them. Don't take my word for it—ask any cat.

I've actually had a glimpse of the Divine Cat myself. It was at Raphaela's first Animal Communication workshop, the one she held in my barn. The cat who showed it to me was Bobbie Kitty herself.

Naturally Christina came to the workshop (Raphaela was right, she did have talent as a communicator), and naturally Bobbie Kitty was with her. There were six people and several companion animals, including, besides the famous Bobbie, an Airedale dog, my own two dogs Julie and Theo, and a chinchilla.

I'd never seen a chinchilla before. They are adorable little rodents, about the same size as guinea pigs, of a soft pearly gray, with the softest, plushest fur you can imagine. The little guy spent most of the time in his cage under his cedar shavings, out of sight of the big Airedale.

We all sat in a circle. The dogs sat with their people, the chinchilla's cage was on a table in the center, and Bobbie was perched on a filing cabinet behind us. She seemed to be observing or perhaps supervising the workshop, even while she remained poised to escape the dogs should it become necessary.

One of my favorite moments of the workshop had just occurred. We were going around the circle practicing communicating with the various animals. I asked the Airedale what he was thinking about, and heard him say, "I'd like to just hold that chinchilla for a minute. I'll give him right back."

I was still laughing about that when my turn came to talk with Bobbie. I glanced over my shoulder to where she was sitting. She looked undeniably regal but not particularly "spiritual," at least as far as I could tell. Then I closed my eyes and tuned in.

Almost immediately my eyes flew open in absolute astonishment. Bobbie had just showed herself to me not as the sleek, flame-colored cat she was but as, truly, a goddess. It was an amazing moment. She was enormous, and filled the space around her with an aura that was not just flame-colored but actually made of flames, like a great burning fire. As my awareness filled with awe and reverence I could hear Bobbie softly laughing at me as she said, "Now do you understand?"

This magnificence, I believe, is the true inner nature of the cat. I haven't had an experience like that one since, but that hasn't done anything to lessen its impact. I believe the only reason I don't witness the Great Cat Goddess more often is because I rarely have clear telepathic experiences unless Raphaela is around.

It's like cello playing. I am an amateur cellist with a passion for chamber music. My husband and I often play string quartets with other amateurs. A few years ago I got to know Emma Rubenstein, the first violinist in the Oxford String Quartet. Emma has let me play quartets with her once or twice, and each time I found myself playing noticeably above my usual level. I guess a little bit of her magic can't help rubbing off on me.

Raphaela does the same thing to my telepathic ability. I doubt very much I would have seen Bobbie Kitty's magnificent aura if she hadn't been there. Still, even though I don't always see cats as the divine beings they are, my experience has definitely affected my attitude. I would even go so far as to say I've become positively Egyptian.

Is It True What They Say About Black Cats?

When I first met Raphaela I had just one cat in my household, a pretty calico named Cocoa. My husband and I thought we were hiring Cocoa as a mouser for the barn. Ha! She soon disabused us of that notion—she never goes near the barn. Only the softest pillows in the house for Lady Cocoa!

Now I have four cats. Besides Cocoa there are Mocha, Butter, and Beta. Butter is a beautiful orange cat. When he first came to live with me I called him Pumpkin, but that was before I asked Raphaela to have a chat with him. He told her, "Pumpkin is all very well, but the *name* is Butter." So Butter he became. Beta is a little black and white tuxedo cat, and her real name is Mehta Beta. "Back in June" is Beta's story.

Mocha is black. He's gorgeous, but even more than that, he has the black cat mystique about him. He got me wondering what it was about black cats. Where does their mystique come from? Why are they always associated with magic? What makes them symbols of Halloween? Why are they invariably pictured either watching a witch stir her cauldron, or riding on her broomstick?

I put the question to Raphaela, who said that as a matter of fact, she did know something about black cats. She had her information directly from the mouth of a black cat named Benoji. Benoji lives with Jasmine Indra, who was one of Raphaela's animal communication students and is now a skilled communicator herself.

As part of their training Raphaela asks her students to do practice consultations with other members of the class. Each student in turn talks to the other students' animals. Jasmine noticed that whenever she was doing one of these training consultations Benoji would wander around in what Jasmine thought was a distracting manner. She'd walk back and forth, swish her tail and hop off and on Jasmine's lap. Once she actually *bit* Jasmine during a consultation, which was totally out of character for her.

Jasmine finally woke up to the fact that Benoji was trying to tell her something. She sat down and tuned in to her, and then asked her what the heck was going on. Benoji told Jasmine, quite forcefully, "Don't you realize I am your helper? Here I am trying my best to help you, and you're not paying attention! You'll be talking to an animal, and I realize you can't hear the animal very well. I can hear perfectly! I could be clarifying the messages for you, if you'd only listen."

Jasmine was excited by Benoji's words, but since she was still a student she didn't totally trust her perception. At the next class session, she asked Raphaela to confirm it for her. "Yes, that's right," Benoji confirmed to Raphaela. "Jasmine heard me just fine. In the old days I would have been called her 'familiar.' It means a helper animal. And yes, everything you've heard about black cats is true!"

Benoji explained to Jasmine that black cats have what she called a "genetic, recessive trait" that enables them to receive information from the natural world and then pass it on to humans who are open to it. This trait was very useful during the Middle Ages, when the Church was in conflict with the older nature religions that still flourished in the countryside. Black cats as a group made the decision to take the wisdom of nature into their bodies and keep it there, so that it would still be available to people who wanted to make use of it.

Benoji showed Jasmine that she and other black cats possess an ability to transmit a sort of "ball" of knowledge, which reminded Jasmine of one of those wooden puzzle balls that children like to take apart and then laboriously put back together. The cat transfers the "ball" directly to a human's mind. When you pick up one of these balls of knowledge from a black cat (which may happen without your quite knowing it has happened), the information feels vaguely puzzling. You have a sense of inner knowingness, without being quite sure what it is you know!

Over time the knowledge given to you by the cat works its way out of the ball and is assimilated into your human thought processes. It clarifies in your mind. You believe you have had some unusual insights. Of course, the whole process works much better if you are able to do it all consciously and have some appreciation of the way your cat is working with you.

Benoji said that not all black cats chose to participate in the project of helping to save nature's wisdom, but enough of them did so to give them the reputation they now have, as the familiars of witches, sorcerers, or shamans.

Black cats as a group have paid a steep price for this generous service to humanity, just as the people who kept alive the ancient wisdom have done. Witches and other such devotees of the old ways have long been objects of persecution, and black cats have been persecuted right along with them. Sadly, they are still sometimes the objects of harassment or abuse. I know I always make sure Mocha is inside on Halloween.

Raphaela learned something more about the way these familiars operate from a very creative black cat named Samantha. Samantha, or Sweet Pea, as she was called, lives with an artist named Dicksie Sandifer in Virginia Beach. Sweet Pea often joins Dicksie when she takes a cup of tea and sits in her studio contemplating the painting she has done the day before. Invariably, new ideas come to her—colors, shapes, ideas for other paintings. Dicksie is a creative person herself, but many of the ideas in her paintings actually originate with Sweet Pea! When Raphaela consulted with her she learned that the cat was helping Dicksie in true black-cat fashion.

Sweet Pea doesn't just help Dicksie with her painting, she also gives her spiritual guidance. Dicksie travels quite a bit, and at one consultation she asked Raphaela to find out the best way for her to keep in touch with Sweet Pea while she was on the road. Sweet Pea replied, "Tell her she needs to talk less, and meditate more!"

Sweet Pea then backed up this excellent advice with practical tips. "Tell her this is the best way to keep in touch. Establish a time, and call me every day at that time. I will be there to tell you how I am and how everyone is. I will take care of everyone. No problem. Nothing could be easier."

Sweet Pea added, "My time on earth is short, and I want to make use of it. My mission in life is to move Dicksie to a new stage in her spiritual development. Every time I meow at her, I am reminding her of that."

Raphaela asked our black-cat advisor Benoji whether he thought artists like Dicksie are now the people most likely to tap into cat wisdom. "Not necessarily," he said. "Sometimes people do think we black cats give special assistance to artists. That's not accurate. We do help a lot of artists with their work, but we don't aggressively target any one group of people. Our wisdom is available to anyone who wants it. All kinds of people are coming to appreciate black cats more fully. Even some normal people, like animal communicators, are starting to value us."

After Jasmine became a professional communicator, Benoji worked actively as her advisor and always sat by her side as she did consultations. One day Jasmine happened to ask Benoji why she bit her during that practice consultation. Benoji said, "That was because you were missing something important from the cat you were communicating with. I needed you to 'get' physically that the cat was in a lot of psychic pain. I bit you so you would understand the cat's pain.

"I don't have to do that any more, because you are better now at listening yourself. Now I help by translating for you when the animal you're talking to isn't clear. Sometimes I also assist you by coming up with a way to help the animal. Do you remember that time I told you and Raphaela what to do with that silly little cat?"

Indeed she did. Benoji was referring to a cat Raphaela had been trying to help for over a year. She had already talked with the cat and her person, Donna, many times. The cat was an unusually small, aristocratic, highly-strung being named Jazz. Jazz was driving Donna crazy with her fixed habit of sitting on a corner of her litter box and peeing over the edge.

When Raphaela asked Jazz why she was doing this, Jazz said it made her feel big! Raphaela's first thought was to try to build up Jazz's self-esteem, so she would feel good about herself and her size. That didn't help—Jazz's self-esteem was just fine, thank you. Raphaela then went on to try every solution she could think of. She asked Donna to change to a deeper litter box. They tried positioning more than one box around the house. They changed the position of the box. They tried different types of litter. They tried flower essences. You name it, they tried it. Nothing helped.

Finally Raphaela decided to bring the problem to her communication class. She asked all the students in the group to tune in to Jazz. Maybe someone would pick up something Raphaela might have missed.

Jasmine was in the class, and as usual Benoji was with her. After everyone had talked with Jazz and come up dry, Benoji whispered to Jasmine, "I know what to do. Smack her. That's what her mother would have done."

Jasmine was totally shocked. The whole practice group was shocked. *Smack* a little cat? They could never suggest that, and of course they didn't. It's one thing for a mother cat to cuff a kitten, quite another for a person to hit an animal.

Everyone stared at Benoji in disapproval, but Benoji held her ground. "I'm right," she insisted. "No mother cat would jump through hoops the way Donna does. We take training our kittens very seriously. This cat is peeing outside the box because she *likes* to. Her mother would have gotten through to her, I can promise you that."

Even though she wasn't about to suggest smacking Jazz, Benoji's suggestion did get Raphaela thinking. The next time she talked with Jazz she said, "You know, your little trick with the litter box has cost Donna hundreds, maybe even thousands of dollars. She's had to pay for all these consultations, not to mention the fortune she's spent for fancy litter. She even had to buy new carpeting, and her landlord is totally on her case because of you. If you were *my* cat, you'd be sleeping in the garage. And I just might tell Donna so."

"Go ahead," said Jazz smugly. "She'll never do it."

That put Raphaela right over the top. "Donna," she said, "If you don't make Jazz sleep in the garage for at least one night, I'm done with her!"

Well, Donna was a pushover for Jazz, but she knew Raphaela was right. She steeled herself to a night of guilt and piteous cries and put Jazz out in the garage. Guess what—it took exactly *one night.* Jazz never peed outside the litter box again.

Thank you, Benoji. You did say black cats are magic!

Back in June

Not long after the workshop where I glimpsed the Divine Cat, I was looking through the Metropolitan Museum of Art Catalogue and I spotted a set of Bastet earrings. There were two little Egyptian cat goddesses, each one dangling from a small gold ring. I had to have them.

The very day I put them on for the first time, two beautiful, healthy, full-grown cats came sauntering up my driveway. They seemed too gorgeous to be strays, but when they were still there a few days later I realized they had come to stay. The black one I called Mocha. The other was a magnificent long-haired tuxedo cat. I named him Mehta because his elegance reminded me of the conductor Zubin Mehta, and conductors are one of the few people who are *always* in their tuxedos.

I called Raphaela and asked her for a consultation to find out who these guys were. She reported, "They're brothers—litter mates, as a matter of fact. They belonged to a student at Humboldt State University who graduated and moved out of the area. I'm getting that the student made some kind of vague arrangements for them—you know, somebody who might feed them from time to time when she thought of it. So these cats thought to themselves, 'Hey—we can do better,' and they took off to seek their fortune. It looks like they've found it with you."

I was convinced it was Bastet who brought them to me, and I was highly satisfied with Raphaela's report. What a great way to get two beautiful new cats!

Then came the problem of integrating them into the household. My only cat at that time was Cocoa, and she was the undisputed ruler of the house. She and my dogs Julie and Theo made up a tight little family group who hung out together during the day and slept on our bed with us at night. I didn't see how I could bring two grown cats into the house with them, especially two big, confident males like Mocha and Mehta. Mehta, especially, was just such an insouciant boulevardier of a cat; his charisma was simply overpowering. I decided the only solution was to have them be outside cats. I would feed them on the porch, and each one would have his own cat condo—a towel-lined box to curl up in and call home.

They accepted my terms and took up residence. Cocoa hated having them, but I thought she'd adjust eventually, especially since Mocha and Mehta weren't going to be inside. But Mehta was never really happy with the arrangement. He'd stand at the sliding glass door leading from the back porch, gazing at me and meowing to come in. The first time I saw him there I went outside and picked him up. He immediately put his arms around my neck, purred, and began touching my nose with his. I think it was at that moment that I fell in love with him.

I actually did. I don't think it's too strong a word, even though I know as a person and a cat Mehta and I had some obvious barriers to becoming an actual couple. Still, it felt the same. I hadn't understood before how much intimacy could exist with a strong, loving cat. I should tell you that I was incredibly flattered, too—I felt like Gigi must have felt when she realized what Maurice Chevalier was actually *thinking*.

I adored everything about him. I loved his faithfulness. While his brother Mocha was a friendly, outgoing cat and made friends easily (maybe the college kid liked him best), Mehta never let anyone else so much as touch him. Also he was unpredictable. Once, while he and I were in each other's arms, he suddenly hit both my cheeks with his paws (claws out) and jumped down. What had I done? He played with me like a heartless lover, or maybe just like a mouse. He was like one of those guys who don't always treat you right, but whom you just can't get enough of.

I thought about him when we were apart, and couldn't wait to get home so I could see him. I promised myself that some day we would spend the night together. My plan was to dream up some excuse why I needed to sleep out in the barn, and then sneak him in with me. I would dwell secretly on this idea, loving it.

There was one dark corner to our relationship. It was that very often when I held Mehta I would feel a premonition of his death. I actually saw him dead. I imagined this thought invaded me because I loved him so much that it made me afraid of how I would bear it when he died. But no, that wasn't it. I felt his death because it wasn't very far away.

One Saturday night in September, I came home and didn't find Mehta on the porch to greet me. He wasn't always there so I went to bed without worrying too much, but the next morning when he didn't come for his breakfast I went out to look for him. His body was just a few steps from my driveway, on the other side of the road. He'd been hit by a car as he tried to cross the horrible, hateful road.

I went to get my husband and somehow got out the news. Together we picked Mehta up and carried him home. We found the perfect spot to bury him, beneath a redwood tree on the hillside behind the barn. Then I called Raphaela, and even though I could hardly speak for shock and grief, I managed to tell her what had happened.

She said, "We'll talk to him."

I struggled to understand. I had never thought it through, but I should have realized that if you can speak telepathically, directly to an animal's heart or mind or soul, the animal doesn't have to be in his body for the communication to take place. When I grasped what Raphaela was saying, part of me did desperately want her to talk to Mehta, but part of me was frightened as well. Mehta was dead. We could never hold each other again. Could I actually bear to be in touch with his soul when his body had been torn from me?

That was part of it. Another part was fear of being gullible, of being led by my pain to succumb to what might be a pathetic delusion. I was also afraid it would somehow be bad for Raphaela. To the extent I thought about it at all, I must have assumed that there is a reason we are divided from the dead. They obviously do pass through a curtain of some kind, and it's thick and strong. I suppose I thought that if we were meant to communicate with them we'd be able to. It wouldn't be esoteric, or hokey either, for that matter. Maybe it was possible, but what if it somehow flouted nature's law? You might pay for that. Who hasn't read stories about people who called up ghosts and then couldn't get rid of them so easily?

All this went through my mind in an instant, but what came out of my mouth was, "Could we really?"

"Of course," Raphaela said calmly. "I do it all the time." Again, just as when she had first told me she was an animal communicator, my mind turned a decisive corner. She'd sounded so perfectly matter-of-fact. Didn't she realize she had just offered proof that the soul survives death? Hadn't she just answered the question that has preoccupied us in our darkest moments since time began? If so, these facts didn't seem to impress her particularly. She just laid it in front of me very naturally. Mehta's soul was still alive. If I wanted her to talk to him, she would do it, just as she had done when he was in his beautiful black and white body. "Yes," I said. "Please."

I waited, holding the phone. Then she came back with a laugh in her voice. "He says that was some accident he was in. He jumped straight out of his body." Thank God, he hadn't suffered. "He says to tell you he knows how hard it is, because of what you meant to each other. He says to tell you he'll be back in June."

I held on to those words during the intense pain of the next few days, and then into the months of fall and winter when grief could strike without warning. As everyone knows who has felt the long knife of grief, you go on because you have to, because life insists on it, and eventually life helps you heal.

All winter June seemed an eternity away. Then finally spring came. Mehta would be coming back. How was he going to find me again? I called Raphaela and asked her for another consultation. This time Mehta said, "Tell Elizabeth not to worry. I found her before. I will find her again." That old swagger of his! I didn't know if I should quite trust it. After all, he thought he could cross the road.

I decided I'd better try and help. I began telling my friends that I was looking for a kitten who would be the reincarnation of Mehta, who said he'd be back in June. That meant he needed to be a newborn kitten about now. Two or three friends had cats to find homes for, and they called me, saying they were pretty sure at least one of their cats was Mehta, if not all of them. But none was a kitten born in September or later, so I knew I hadn't found him yet.

June came. I went through the month calling the Humane Society every couple of days to see if any kittens had been brought in for adoption. "No kittens today," I was told time after time. "You just missed them. We had some this morning. Call back, check again."

Raphaela came to town, accompanied by Mary Getten, for a weekend of work on the orca interviews. I asked them both to tune in to Mehta. They both got a message that he was at the Humane Society, in the body of a gray and white cat. We drove over there, and there was a gray and white cat but he was too old to be Mehta. I don't know quite how to explain this except to guess that when they called Mehta, this cat somehow got on the line. Maybe wrong numbers can happen in telepathy, just like in telephony. Who knows?

Then, the last week in June, my hairdresser Kari told me about a friend of hers, Freddie Fleanor, who was trying to find homes for a litter of kittens born wild under her house. It was the next weekend before I was able to drive over there, by which time Freddie had been able to bring three of the kittens into the house. She showed me the tamest of them, a darling little gray. He was precious, and I was more than ready for another cat by now. But was this Mehta?

Then I spotted another kitten under the couch. It was a perfect tiny black and white tuxedo kitten. My heart turned over. It *was* Mehta. It had to be. Then a sudden thought made me hesitate. Mehta had promised me he'd come back in June. Today's date was July 1st. Oh, never mind! I thought. Who cares! "That's the one," I told Freddie firmly. "I'll take him."

Freddie was happy I was taking the kitten, although she would have been happier if I'd taken the gray one too. "This is the second litter I've had to find homes for," she said. "I've tried and tried to catch the mother to have her spayed, but I never can."

I clutched the little kitten to me, already more than half in love. The name came to me: Mehta Beta. Then Bastet, the Divine Cat, who was living inside the trembling little body in my hands as she lives in all cats, opened the door of the future just a crack, and let me glimpse the truth of our life together.

I saw that you can never have the same cat twice. That this little cat was going to be unspeakably precious to me. That I would never, ever let her go outside. That she would never grow very big, and that I would never call her Mehta Beta, and most of the time not even Beta. I would call her Baby.

I saw that she was Mehta, and at the same time she was Baby and no one else. And so it has come about. She's Mehta in the way she purrs, loudly, nonstop. She's Baby in the way she gently licks my hand when I hold it out to her. She's Mehta in the way she tumbles around with Mocha—exactly the way Mehta used to, and right from the first moment, as if they recognized each other. She's Baby in the way that, yes, she's a girl. She's my darling come back to me, and she's her own dear self. I still miss Mehta, and I couldn't possibly love Baby more.

"Thank you, Freddie," I said. "I'll take good care of her. I wish I could tell her mother so. What's her mother's name, by the way?"

"The mother cat?" said Freddie, "Oh, I call her June."

Lisa: If They Can't Hear Me, I Must Speak

Lisa sat quietly in the boathouse, patiently waiting for George and Harriet to come and get her. She had perfect faith that they would, but she was starting to wonder why they didn't hurry up.

They'd already been in the boathouse twice—they just hadn't looked behind the broken window into this dusty little corner room. Probably they didn't imagine she could be in here. And she wouldn't be, either, except that a monster had been chasing her (dog, raccoon, bear—Lisa didn't knew what it was and didn't care), and she'd raced up a trellis and almost flown through a half-broken pane of glass. Now she couldn't get out or even imagine how she might get out if George and Harriet didn't come.

The second time they'd been in here looking she noticed they'd looked terribly upset. She could hear their frantic thoughts. She'd sent her own thoughts out in response, telling them that she hadn't eaten in three days, that she was hot and thirsty, that she was ready to go home.

Lisa knew she meant the world to George and Harriet. Of course they'd be combing the neighborhood, knocking on doors, doing everything they could to find her. Her picture would be on every tree. Why didn't they do the one simple thing, and tune in to her thoughts so they could find her?

Why indeed? It seems incredible that they wouldn't know how easily they could have communicated telepathically with Lisa. But they didn't, and they might never have learned any different if George and Harriet's daughter, who lived in San Francisco, hadn't called Raphaela and begged her to try to find her parents' cat in Alabama.

Raphaela was much more experienced by this time than when she'd taken on the case of Bobbie Kitty. She fully understood by this time how difficult it was to find lost animals; and in this case she didn't even know if she'd be able to contact Lisa. Lisa wasn't even her client's cat! But something told her to go ahead and try.

To her surprise Lisa answered her call promptly and clearly. "A remarkable cat," Raphaela noted. Immediately she could feel the muggy, dusty place. Lisa showed her the monster that had chased her, then showed how she had escaped by racing up a trellis and leaping into this shed. Raphaela thought it looked like a garden shed. Lisa then showed Raphaela how she had peered out through cracks in the board floor and had seen George and Harriet—twice!

"For goodness sake," Raphaela told her, "when they come again don't just sit there! Speak up, cat! Make some noise! Yell! Scratch! Do whatever you have to do—just get their attention!"

"Well, okay, if you say so," said Lisa. She sounded surprised but willing. Raphaela didn't think Lisa had any idea why such a crude form of communication would be necessary, but she took a sensible attitude—"I'll try anything"—the way you might respond if your doctor suggested something silly but harmless, like soaking your feet in water to cure a cold. If you're desperate enough, you'll give it a try.

Twelve hours later, George and Harriet, who had searched every garden shed in the neighborhood, walked into the boathouse for the third time. They told Raphaela they'd looked up in amazement to see their Lisa, who was normally a rather silent cat, yowling, pacing, and peering down at them as she screeched like a mountain lion.

Sometimes I wonder if our cats have too much faith in us. Lisa still doesn't understand why she had to give up her silent dignity. After all, George and Harriet had been divining her every wish, and granting it, for her whole life. How was she to have known they couldn't hear her thoughts, as she heard theirs? But at least Lisa was willing to come down to our noisy human level when it was really important!

Mona, the Yogic Cat

If you'd like to set an impossible task for yourself, try to make yourself worthy of your cat.

Of course you'll never succeed. But if you have a cat that's truly an enlightened being, and is willing to help you on your spiritual path, shouldn't you at least *try* to master the lessons your cat provides to you?

Anyway, that's the way Mona, a sleek and energetic white cat, saw things. Raphaela's client saw them rather differently. Deborah thought of herself as a big-hearted, beneficent earth mother. She loved to rescue cats and add them to her cat family; she had already brought three cats and a large bird into the house. Mona, on the other hand, saw Deborah as a rather disorganized, impulsive person who needed to settle down, pay attention, and focus on Mona. *Then*, she felt, Deborah might begin to make some personal progress.

Deborah had consulted Raphaela because Mona had resorted to that all-purpose cat method of getting a person's attention, peeing outside of the box. When Raphaela tuned in to Mona and asked her why she was misbehaving, Mona told her, "I'm doing it for a reason. I want you to tell Deborah that she shouldn't bring any more cats into the house." The litter box was already a mess, and Deborah wasn't that great about cleaning it up. Mona would go back to using it, but only if Deborah made a few changes in her *modus operandi.*

Raphaela's sympathies settled with the cat. She agreed with Mona that rescuing animals is fine, but a person needs to make a continuing commitment to a rescued animal. Deborah's commitment to follow-through struck Raphaela as being rather limp. And in fact, this turned out to be typical of Deborah generally. She was the kind of person who likes to dabble in spiritual practices, but didn't concentrate on anything for very long. She'd already been into aromatherapy, Reiki healing, and who knows what else—and moved on from each one.

After several consultations with Deborah and Mona, Raphaela began to feel a bit exasperated. At this point she happened on a verse in the *Bhagavad Gita* that deals with people who don't follow through on their spiritual practices.

> *"He who strays from the path of Yoga, after ten thousand years is reborn in the house of the pure and illustrious—or else into an actual family of Yogis, although such a birth is more difficult to attain."*

The verse improved her perspective. She didn't need to be judgmental about Deborah's path. Let her dabble—maybe that was what this lifetime was about for her. She had the backing of an industrious, supportive husband who seemed perfectly content to have his life revolve around Deborah and her practice of the moment. No one's personal evolution was on anyone else's time line, after all, and there didn't seem to be a hurry. Eternity was going to be there. Besides, Deborah had Mona!

After a few more consultations, Deborah decided to join one of Raphaela's communication classes. Perhaps being an animal communicator was her role in life. Raphaela thought it probably wasn't, but having Mona in the class turned out to be a truly amazing experience for everyone.

At one of the first class sessions, Mona said, "Leave Deborah to me. I can help her!' She then offered to spend fifteen minutes a day sitting with Deborah doing Zazen meditation. Next she taught Deborah a Yogic breathing technique, a simple practice that the whole class was soon using. What didn't this cat know?

Deborah dropped out of the communication class not long afterwards, but Raphaela wasn't worried. She was sure that Deborah's wandering ways were over, and that she would be absolutely sure to stick with this meditation practice. After all, how many people have their own enlightened cat to sit with them every day and guide them on their way? Maybe this is what happens when you stray from the path of Yoga, she thought. You are reborn into an actual family of yogic *cats*.

A few months later, Raphaela checked in with Deborah to see how her meditations with Mona were going. "Oh," Deborah said, "I'm not doing that any more."

Why did she stop? She couldn't say, and neither could Mona. But now you know why cats sometimes look at us with incredible patience in their eyes. Eternity is a long time, they seem to be saying, and we can wait.

Max and Dee

"Sometimes I feel such a connection with my clients and their companion animals," Raphaela told me one day. "I feel that way about Dee and her cat, Max. Dee is just a wonderful woman. She's been meditating faithfully for more than fifty years, and you can just feel the depth of silence in her. She's an artist, mostly retired, and she lives with two long-haired Dachshunds and four cats.

"Dee and I had talked many times and I'd communicated with all of her animals. They all adored her. It was a truly happy family. Almost a year ago, I learned that Dee was facing what all happy families face eventually. Max, her twelve-year-old gray tabby, was ill. He had been diagnosed with mast cell tumors, a kind of skin cancer. Dee had had Max's tumors removed, but they had come back.

"The reason for Dee's call was to ask me to find out from Max what he wanted her to do. He could have more surgery, or they could try chemotherapy or radiation. I tuned in to Max, and he said, 'No. Do nothing. I'm happy as I am. My appetite is fine, I like to eat and lie in the sun, I enjoy the other animals.'

"Dee accepted Max's decision. She asked me to check in with him regularly through the months that followed, and he always said the same thing: I'm happy, I like to eat and lie in the sun, I enjoy the other animals. He was like a record.

"Ten months after the first consultation, Dee called me again. This time she said, 'I don't know what to do. Max looks like hell. He's covered with tumors. I don't care what he says, he has to be suffering. I know I am.'

"This time when I tuned in to Max, he did sound different. He still didn't quite admit he was sick—'I'm not comfortable' was as far as he would go, but that was farther than he'd gone before. He said he knew that it was tough for Dee and that the other animals were upset by his condition. He understood that Dee was offering him euthanasia if he wanted it, but he said it wasn't time yet. He said, 'Tell Dee I'll let her know when I'm ready.'

"Dee of course said, 'Ask him how he'll tell me.'

"'I'll sit on the coffee table in the living room. That will be your sign that I'm ready,' Max said.

"'That must be right,' Dee said, 'Because Max never sits on that table. Tell him I'll respect his decision. I leave it completely in his hands.'

"Five days later, Dee called. She said that in the living room is a special chair where she always does her meditation. That morning she went into the living room and sat down in her chair. Max followed her and then climbed up onto the coffee table. Dee looked at him, understood, and then closed her eyes. For an hour she sat meditating with her beloved friend. Once or twice she opened her eyes to look at Max, and he each time he was there, sitting quietly, gazing back at her in perfect peace.

"It was such a blessing to hear this story," Raphaela said. "I feel privileged to have played a small role in helping these loving friends say good bye to one other in as much peace and harmony as they lived. This is why I became an animal communicator. I only hope that when I die, the Divine Cat will help me do so with half the dignity, serenity, and compassion of Max."

Chapter Two

God Spelled Backwards

Theo: The Soul of Tibet

Snow was falling in great, soft flakes outside the monastery high in the mountains of Tibet. Inside, a monk lay dying. Just before he closed his eyes and released his final breath, his eyes traveled around the Great Hall, lit by the flickering of many butter lamps. All his fellow monks were there, chanting quietly, keeping him company to the gates of the other world. His last thought was of all the kindness he had known in this lifetime.

After his death the lama spent some time floating between worlds, enjoying the calm peacefulness of being free of a body. He knew he had not stepped completely off the Great Wheel, but he hoped he would be able to do so after another lifetime or two in the lamasery where he had spent these last many incarnations.

He remembered how it had all begun. He had been a simple peasant, living on a small farm in the valley, almost in the shadow of the great lamasery perched near the summit. Times were hard. His father had said, "My son, you must go to the monks. There you can eat. Here, I think you cannot." That first life as a monk had been difficult. He remembered chanting, extreme cold, and very little food.

Yet the spiritual life had gradually taken hold of him. In this last lifetime he had joined the monks of his own free will when he was still a boy. The life was not different—there was still chanting, cold, and very little food, but now what he remembered was the devotion, steadfastness, and purity of the life, and not the very thin clear soup that made up most of their meals.

A few more lives spent practicing compassion and non-attachment, and in doing his work as a stonecutter for the monastery, should bring him to enlightenment and the fulfillment of his evolution. Meanwhile, he rested. Then rebirth came as it always does, in a rush of sensation and forgetting. He passed several lifetimes away from Tibet. Then, to his great relief, he found himself once more outside his monastery, awaiting rebirth. He thought he was on his way back to the life he loved best when he slipped into the body that awaited him.

When he awoke in the new body, he wasn't able to make out where he was. Something smelled different. There was the delicious aroma of milk, that was normal at the start of a new life, but where was the wood smoke and sandal paste and the homey smell of a nearby yak, the well-loved smells of a traditional Tibetan home?

Then he felt an unfamiliar sensation: a rough, warm tongue was licking his body. It was curiously invigorating, and he moved his limbs in an experimental way. There was something very different this time. Four little paws, instead of hands and feet! A tail! Fur all over! The last moments before incarnation came back to him. At the gateway of life and death, where he always waited for the gesture that would send him into rebirth in a home near his lamasery, the angel at the gate had pointed in a different direction. His gaze had traveled down, down from the high mountains of Tibet, across the plains, across an unimaginably huge ocean to the edge of another continent. "A lifetime of love and service," the angel had whispered into his ear as he set off in the direction. He hadn't understood, for were not all his lifetimes given to love and service?

Now, as he pondered these matters, he felt his body being picked up and dangled in a harsh light. "Nice," someone said. "A perfect male Shi' Tzu. We'll keep him as a breeder." Then he felt himself being dropped back next to his mother. He wriggled luxuriously in her warmth, making his way toward her milk.

He remembered something he had often heard his head lama say. Every being in the world deserves our love and compassion, because any one of them may have been our mother in another life. Now his mother was a lovely little Shi' Tzu dog, red and white like him. He whispered to her that he had come from Tibet. "We are all from there, dear," she said. "We're a long way from home. You'll see soon enough, my poor little one."

Soon he understood her sadness. The period of closeness with his mother lasted only a short time. Then he was taken away from her, and his life filled with loneliness. He was placed in a small cage, removed from loving contact with people or other dogs. It was a life of poverty and deprivation—not the cheerful austerity he had known in Tibet, but a life of coldness, without love, direction, or prayer. He could hear and see other dogs around him, but none of them could run, play, or share warmth. Worst of all, he almost never touched the ground. How he missed the feel of the earth, the real earth, beneath his feet.

Being the contemplative soul that he was, he settled down to pass the time in meditation; but it did not feel right. He found he missed the snow on the mountains. The memory of snowy mornings, when he would go out the little door in the kitchen to fetch wood for the cooking fire, was precious to him, and he replayed it in his mind over and over.

When he became an adult, he became part of the breeding routine. This was a completely unexpected part of his new life; he had been a monk for so many lifetimes he had almost forgotten about these particular pleasures of the flesh. This was nothing like what he remembered from his last householder life, generations ago. Where was the companionship and intimacy of marriage? Where were the children? He never saw them, though he knew they must exist. Each time he was taken from his cage to be bred, he heaved a deep, Tibetan sigh, part indignation, part resignation.

He did not know how many years went by in this way. He never saw either the sun or the snow, so there was nothing by which he could tell one season from another. Lights were kept burning at all times, so even the cycle of day and night was denied to him. The routine was not pleasant. The food was scanty and brought to him without the least personal interest. He was bored, often hungry and always lonely, but it was all he knew.

Then one day there was complete disruption. Strange people came into his area, speaking roughly and loudly. A man opened his cage, took him and held him up for inspection. "Just look at this poor little guy," someone else said. "We'll take him as evidence. I'd like to shoot the people who do this to dogs."

Later the little Shi' Tzu could barely remember the dizzying days that followed. He was handed from person to person—more people than he had seen in his entire life. One day, frighteningly, his stomach was tattooed with an identification number, and he became an official piece of Los Angeles County evidence. Then he was left for days in a small cage. There were cages on all sides of his, filled with other lonely, disconsolate dogs. He listened carefully to the people who moved in and out of the room, hoping for some word about what would happen to him. Was this the time for him to begin the life of compassion and service he had entered this body for?

He thought he picked up the word "euthanasia." It was a strange word, but he sensed what it meant: the end of this incarnation. Then, to his surprise he heard a new voice, filled with more kindness than he had yet heard. A young woman took him from his cage and held him close. "His last day?" she said. "Thank you for telling me. I'm from the Lange Foundation, and we will take any dog who's having his last day. This one's darling. We'll certainly be able to find a home for him." The nameless little dog, who had been a Tibetan lama and then a breeding dog in a hateful puppy mill, was about to become my Theo.

I met Theo in the office of a kindly veterinarian who kept dogs rescued by Gillian Lange and her Foundation until a family could be found for them. At the time I went to choose a dog there were about a dozen rescued animals living there. I saw the little Shi' Tzu with the mysterious air of dignified reserve, and immediately asked if I could take him home.

We had been told about his history, and looked forward to giving him a new and happier life. We soon found that his restricted life in the puppy mill had left him unequipped for the ways of a family home. Stairs baffled him. He had never walked more than a few steps, and had trouble doing it; he seemed to prefer walking in a circle, as he would have in his cage. He was afraid to go outside, especially when it was dark. He barked wildly at any stranger, though he always kept a safe distance away.

How could we help the little guy learn to be a real dog? How could we teach him to relax and play? We thought about it, then went back to the Lange Foundation and brought home a five-month-old puppy. Julie was a Llasa-poodle cross with dark gray curly hair and little black button eyes. She was about Theo's size, 12 pounds or so. She had been abandoned by her people, but she bore no scars and seemed to be a completely normal puppy.

Within a few hours Julie had explored the entire house, claimed it as her territory, and begun her mission: showing Theo how to live. For his part he accepted her immediately as his teacher. We understood clearly that in Theo's pantheon Julie stood first. Then came my husband and me, distant seconds.

Most of the time Julie merely bossed Theo around, going through every door first, snatching the remains of his dog biscuit when she had finished her own, and leading him astray in the matter of our neighbor's cat. However, if she got upset she'd turn on him and briefly, very briefly beat him up. It never lasted more than a few seconds and she never hurt him, but it looked pretty wild from the outside. Theo, true to his Tibetan Buddhist background, always maintained a strict nonviolence policy. He never responded to Julie's outbursts in any way, just bowed his head quietly and waited until it was over. His devotion to her never wavered.

When Raphaela came to Eureka for the first time she opened up the whole world of telepathic communication to me. I was thrilled with the idea of learning more about my friends Julie and Theo, and I still have the notes from that first conversation on August 31, 1994.

Raphaela began by asking me what questions I had for them. I asked her to ask Julie what she thought about Theo. She said, "Theo is my younger brother. I love him. I boss him. I take care of him. He is in my care. I have a lot of responsibility."

"I have a much larger spirit than this small body shows," Julie went on. "I have been a panther and a hunter in past lives. Now I have chosen to accompany Elizabeth and Ralph on this life, to help them and their music. They think I'm cute and fuzzy but I have a serious purpose. I am a channel to send the music out to the natural world. It is a gift.

"Music and being with Elizabeth are the most important goals in this life. I assist her with my concentration and attention. When Elizabeth plays the cello I go into a meditative state that creates a space for the music to flow outward and upward. Cocoa can come in while Elizabeth is playing, but she is not allowed to distract her. Music is not a game. It's serious."

Well! And to think I had imagined she was just waiting for a dog biscuit!

Raphaela asked Theo about music too. He said, "I feel it inwardly. It is personal. I like the cello. It is rich and somber. The violin is more ethereal. Together they are good. Music soothes and feeds me."

I wanted to know whether Theo was still feeling the effect of his past sufferings in the puppy mill. His reply was so sweet and loving that I often think of it when trying to deal with some tribulation of my own. He said, "I am not much affected by the past. We are all products of our past and our experiences, but I do not dwell on it. I am happy now, safe and secure. I have friends and companionship. Who could ask for more?"

Who indeed? I too have friends and companionship—Julie and Theo first among them. What is a tribulation when I have them?

Theo didn't tell Raphaela about his Tibetan past in their first conversation; it came up several conversations later. When I learned of it I was intrigued but not tremendously surprised. There had always been something so calm, wise, and accepting about Theo that I had no trouble visualizing him in that role.

He spoke with Raphaela about his lifetimes as a monk quite freely and told her many details life in the lamasery. In his last lifetime there, when he joined the monks as a young boy, he had kept close ties with his family on their nearby farm. He often returned to his family, especially during festivals and at harvest, to help with the animals and the farm. The monastery was very much a part of the community. Many of the young men from the neighborhood became monks. It was a life of spirituality but not of withdrawal from the world. Instead, the world was drawn into the monks' spiritual sphere, so that the lives of the householders and the monks grew and evolved together.

In their most recent conversation, Raphaela asked Theo about the lessons of his Tibetan lifetime. He spoke of devotion, steadfastness, and purity (qualities he has shown to us from his first day in our home). He added, "It's to be satisfied with little, and make scarcity rich."

Raphaela said she couldn't help asking how he thought it happened that he, an important monk in Tibet, had ended up in a puppy mill in California. He explained, "I didn't understand it myself at first, but I've meditated about it and it has become clear at last. It was a rather unusual situation.

"You probably know that my kind of dog, the Shi' Tzu, comes from Tibet. We are related to the kind of dog you call Pekinese. We just call them Chinese dogs. They are cousins of ours, but we Shi' Tzus are rougher and hardier, better suited for cold weather and high mountains. You may have noticed that I'm a strong, hardy little dog."

Raphaela made suitably admiring noises, and Theo went on, "Of course, every Shi' Tzu must have a Tibetan soul. There have always been plenty of souls for the Shi' Tzus in Tibet, but this idea of a "puppy mill" for Shi' Tzus meant there were suddenly many more dogs who needed Tibetan souls. Many souls, including mine, were called to America to go into the bodies of all these puppies from the mill.

"A puppy mill is not a good place. I've already told you that my life there was difficult; I am used to austerity but this was not the right kind, this was not good for the soul. Still, we are taught to find blessings in everything. We Tibetans saw that if we went into Shi' Tzu bodies, we could bring the teaching of Buddha to America.

"What is the teaching of Buddha? It is to be kind, loving, humble, and compassionate. We Shi' Tzus try to be good Buddhists. We go among you as humble dogs. We are small. We threaten no one. We bring peace. This is the role we play, and it is one I have been happy to take on. And now—look what happened!"

"What do you mean? What has happened?" asked Raphaela.

"My sufferings have ended, and I have landed in a wonderful life! I am with my friends Julie, Ralph, and Elizabeth. I live in luxury. Perhaps it is unsuitable for a monk, but I find I can get used to it. I sleep on a bed, I ride in a car. Our household is happy and harmonious, thanks partly to me. And for once in all my lifetimes, I have plenty of food!"

When I heard about all this from Raphaela I realized how much Theo resembled one of my favorite characters in literature: Teshoo Lama in *Kim*, Rudyard Kipling's great story of espionage and enlightenment. Theo and Teshoo share an inward quality, a calm, grave concern for others, and an attentive interest in having their food bowl filled. I can almost hear Theo say, like Teshoo, "Just is the Wheel, swerving not a hair." Even their names are similar, and I like to think that perhaps Theo was called Teshoo when he was at home in his lamasery. Today Theo is lying on his bed, regarding me with eyes full of love, compassion, and repose. He has eaten well this morning. I want him to enjoy every single minute that remains to him of this luxurious life in America. But I wonder—should I get him some thin, clear soup? Or perhaps he would like a butter lamp?

Raphaela's Journal, December 1998: The Code of Ethics

I feel that I need to write about what happened tonight with Corrine and Carly. It was one of my saddest consultations, but I hope to put it to at least some use by clarifying my own values.

Penelope Smith has proposed a Code of Ethics for animal communicators. I respect Penelope a great deal, and have always take the Code of Ethics very seriously. However, it isn't an absolute. There are times when I need to listen to my own feelings about what an animal needs. This was one of those times.

The part of the Code of Ethics I'm thinking of concerns the client's value system. It states, "We go only where we are asked to help, so that others are receptive and we truly can help." Another key sentence reads, "We let clients decide for themselves how to work with healing their animal companion's distress, disease, or injury, given all the information available."

That's great up to a point. If a client is comfortable with the idea of reincarnation, I'm more than happy to ask the animal about his or her past lives. If the client's beliefs don't extend that far, I won't go there with the animal. If the dog has been chewing the rug, we talk about physical solutions. What's the use of saying that the dog chews because she was a hamster in her last life?

It usually works out well because there is almost always a good match between the client and the animal. Spiritually inclined clients have spiritual dogs; show clients have dogs that like to be in shows, clients who like the outdoors have dogs who love to be outside. Or maybe it's just that dogs are very willing to go along with thc client's way of looking at the world. If the client is more open, the dog will share more of its spiritual side; otherwise, he'll keep that part of his life to himself.

Where I have trouble with the Code of Ethics is when the client's belief system is actually harming the animal. Fortunately, this is extremely rare, but it was happening in the case of a client named Corrine. Corrine had two dogs, and they were both having health problems. The thing was, Corrine belonged to a religion that didn't believe in medicine. She believed that illness was unreal, a kind of mistake, and that a sick person would be fine if she could realize this fact fully.

Corrine was sincerely committed to keeping her animals healthy, and even made them homemade healthy dog biscuits. Unfortunately, her ideas of what was good for her dogs didn't agree with current nutritional thinking. She kept them on a low-fat diet, which may be healthy for people but not for dogs. Both her dogs were underweight. One of them, a charming little Llasa named Josie, was an insulin-dependent diabetic.

It was really hard for Corrine to accept the chronic nature of diabetes and the ongoing need for medication. In a number of consultations, Corrine explored the idea of decreasing Josie's insulin dose. She explained that she felt this was okay because Josie's body could now make her own insulin again. Corrine had been assured by a psychic energy worker that Josie's pancreas was healed and functioning normally.

I urged Corrine to keep working with her regular vet, monitoring Josie's glucose. If her glucose level stayed normal on a reduced insulin dose, great. But if not, it would be really dangerous to stop the insulin. I hesitated to push my point of view too strongly, worried that I would be imposing my judgment and invalidating Corrine's. But how could I let Corrine decide for herself how to handle Josie's diabetes if her decision involved throwing Josie into hypoglycemia?

I liked Corrine and I respected her metaphysical beliefs, but they were so different from mine. I felt confused, and I tried to make it okay by ignoring the conflict.

Then Corrine called me about Carly, her other dog, a Boxer mix. She told me that Carly had been vomiting intermittently for days. Carly was already thin, but now she was extremely thin and very dehydrated. When I tuned in to her I was alarmed. Carly complained of severe abdominal pains, weakness and nausea. She was thirsty but couldn't keep any water down. I felt the situation was urgent, and I told Corrine to take Carly to the vet immediately.

Corrine was waiting for an appointment in the office of Cheryl Schwartz, DVM, a vet who specializes in alternative medicine. I'm an admirer of Dr. Schwartz, too, but she's very popular and Corrine's appointment was still ten days away.

This was my dilemma. How far should I go in confronting Corrine? It wasn't that she was neglecting her dogs; she was doing her best according to her own belief system, but that belief system was different from mine. I'd already told Corrine that Carly really needed a vet. I decided I couldn't push it, but I felt extremely uneasy.

Finally the day arrived for Carly's appointment. There was nothing "alternative" about the vet's reaction: she was shocked and aghast. She told Corrine that Carly didn't need acupuncture or homeopathy, she needed the best Western medicine had to offer and she needed it right now. She sent Corrine right off to the emergency room, saying that Carly was badly dehydrated and should be placed on IV fluids. She needed a thorough workup, including an ultrasound for a possible abdominal tumor.

Carly left Dr. Schwartz' office but didn't go to an emergency room. She also didn't dare take Carly back to her regular vet. I found out later that it was because she was afraid of what her own vet would say about Josie, whom she had actually taken off insulin. She cast around and finally found yet another vet who would see Carly on an emergency basis. This vet did put Carly on IV fluids and asked to keep her overnight. Corrine refused and took Carly home. That night, Carly died.

Corrine called me, sobbing, in the middle of the night. She told me everything that had happened since our previous consultation. I was horrified. I felt so sorry for Carly, and guilty that I hadn't been more forceful about how desperately she needed treatment. I didn't know whether she could have been saved ultimately—she had certainly been very ill—but I do know that with good medical treatment and a nourishing, high-calorie diet she could have lived much longer and been much more comfortable, instead of dying in misery as she had.

Corrine asked me to contact Carly one more time. I agreed to try, more for Carly's sake than for Corrine's. I found that Carly didn't condemn Corrine. Dogs just aren't like that. She accepted what had happened, but she was a little remote. She didn't cover Corrine with messages of love, as dogs who have recently died usually do.

I apologized to Carly personally, saying that I wished I had advocated more strongly for proper medical care. She refused to reproach me; she just said, "I don't think she would have taken me anyway."

I am always humbled by the forgiveness that dogs show us. It is extremely beautiful, but I don't want to use it to gloss over my own mistakes. If I don't learn from my mistakes, I'm not worthy of Carly's forgiveness.

I have resolved that it is absolutely not okay to close my eyes to an animal's suffering in order not to put a client on the spot. No matter what the Code of Ethics say, I can't use a client's belief system as an excuse to avoid my own responsibility. Yes, if I'd told Corrine everything I was thinking it might have alienated her. There were several other animal communicators Corrine wouldn't work with because they confronted her actions too firmly. That shouldn't matter. My responsibility isn't to keep a client happy at all costs; my responsibility is to be an advocate for the best interests of the animal, to the very best of my ability.

The animal is actually my client, if I really think about it. The animal's belief system is what matters. No, that's not even it. What matters is the animal's life.

Noor: A Little Life

Sharon heard the whispers even before she was fully awake. "Must the children really go? Is it that bad?" That was her mother's voice, sounding terribly anxious. Her father whispered back, "It's that bad. The order just came in—evacuate all the Embassy children. Get Sharon dressed, and hurry. Hurry!"

The next thing Sharon knew, she was being thrust into the back seat of a car. It was still dark, but she could see the glow of fires in the distance and hear loud, upsetting sounds, like fireworks. Suddenly she cried out, "We have to go back! I have to get Noor!"

"There's no time to worry about a dog now," her father said roughly. The door slammed and the car sped Sharon away through the night.

Back in the Embassy, Noor too was wide awake. The little black and white dog knew something was terribly wrong. She usually slept at the foot of Sharon's bed, but knew better than to stay there when big people came into the room. From her hiding place beneath the bed she had watched as Sharon's mother hurriedly dressed her, then picked her up and carried away. Noor couldn't understand it. It was dark, the middle of the night. Where could she possibly be going?

All the next day Noor stayed under Sharon's bed, trying to ignore the noise and confusion that filled the Embassy compound.

Through the constant ringing of the telephone, the sound of hurrying footsteps, and the many voices filled with stress and tension, she listened for Sharon's soft voice and eager running steps. In her heart she carefully guarded the memory of the last thing she had heard telepathically from Sharon—" We have to go back! I have to get Noor!" "This is the last place she saw me," Noor reasoned. "When she comes looking, this is where I must be."

Hunger drove her out at dusk the next day. The little dog crept into the kitchen, where Sharon always fed her. She had never yet forgotten Noor's dinner, so probably she would be waiting for her at dinner time.

A sharp kick sent her reeling against the kitchen door. She scurried into a corner, then waited for a chance to race quickly back to the bedroom. Once again safely under the bed, Noor licked her sore side and began to think. The kick had revived memories of her life on the street.

Before Sharon brought her home she had never had a human friend. Her friends were other dogs; there were three she remembered being close to, but friendship is always difficult when you are hungry and miserable most of the time. Her own little group, small black and white creatures like herself, worked together to scavenge bits of food, and had slept together in a back alley for warmth. Other groups were not so friendly, and would chase her if she approached their chosen garbage spots, but she could hardly blame them for that. Everyone had to make a living as best he could, she knew that.

People were distant presences. A few were kind and had thrown her and her friends a scrap once in a while. Most were indifferent; some were hostile. She had heard herself called "unclean" and had been the object of several kicks and blows. She decided not to think more about that right now.

Noor's nature was soft, loving, and forgiving. She had a vast reservoir of affection in her heart, and would have given it freely if anyone had asked her for it. But she had never dreamed, in the remotest stretch of her imagination, of someone like Sharon. She hadn't even understood at first what Sharon was. In her city the women and girls stayed mostly inside, but Sharon and her mother walked the streets freely, like men. The small girl had called out to her one day as she walked by. "Come here, little dog," she said. Noor was not accustomed to approaching people, so she hid. The next day the girl walked by again and held out her hand to Noor. She smelled food, and approached cautiously, ready to bolt, but it was true: there was a delicious bit of bread in the girl's hand. She snatched it and ran a short ways away to eat, but kept her ears open, listening.

"Look, Mommy, isn't she sweet?"

"She's not very clean."

"I'd wash her."

"Sharon, no. People here don't have dogs for pets. Daddy has his position to think about."

"Please, Mommy."

"We'll see."

This was all very puzzling, but what happened a few days later was positively mind-boggling. The little girl returned, accompanied this time by a tall man with a kind, remote voice. "Oh, I suppose you may," he said, and then Noor found herself lifted up, held well away from the man's body, and carried through a nearby gate.

There she was washed, not once but twice—first outside, with cold water, then inside, with hot water and soap. The strange feelings and smells might have alarmed her, but she was constantly reassured and warmed by the love she felt from the little girl, whose name, she discovered, was Sharon.

For almost two years now she and Sharon had lived together in what Noor thought was probably Paradise. They were never separated. When Sharon had her lessons, Noor sat beside her chair and learned too. They played together in the garden of the compound, and slept together on their own clean, soft bed. Now, from her hiding place beneath the bed, Noor considered what she should do. It was almost dark, and so she decided she would sleep, even though she hadn't eaten. Then she'd think again in the morning.

All that night Noor felt Sharon's thoughts on her. Her mind was always open to Sharon's, and the fact that they were no longer together physically didn't alter that in the least. She knew Sharon was distressed and lonely, and that she, Noor, could help her. She was needed. In the morning, without any more thought, Noor left the bed and went to look for her friend.

The familiar gate of the compound was open, and in the early morning light the streets were quiet. She slipped out the gate and began trotting along, keeping close to the buildings and well out of the street. She had no real plan. She would just go up and down the streets until she found Sharon. She did not doubt that when she came to the place where Sharon was, she would know it.

This became Noor's life, though it was not easy for her. Even staying alive was hard. She had forgotten much of her street wisdom and could not find her friends. Scraps were hard to come by for a dog alone. The first time she tried eating camel's dung, she doubted it would be able to sustain her; but it was her only choice.

She became steadily weaker as the weeks went by, yet she was never discouraged nor disheartened. Every night before she went to sleep she felt Sharon's love for her, as strong as it had ever been. Each morning she thought, "Today I will find her."

That was her last thought, too, before she left her starved body behind in an alley and followed her soul to God.

Twenty years went by. Sharon finished school, married, and had a daughter of her own. She did not forget the little dog she had been forced to leave behind, but unlike Noor, she was not at peace. On the contrary, she was afraid Noor was angry at her for abandoning her. When a friend told her about Raphaela's ability to talk with animals even after they had left their bodies, she called and asked whether it would still be possible to communicate if the dog who had been dead for many years.

Raphaela wasn't sure, but she remembered her own dog Petey, who spoke to her telepathically even after he'd found a new body as a hawk. She told Sharon she would try. "Noor may have been in several other bodies since she was with you," she told Sharon, "but that doesn't mean she won't remember when she was Noor."

Noor responded to Raphaela's call. Yes, she still remembered that incarnation and she certainly remembered Sharon.

"Ask her if she's angry at me."

"Oh, no!" the little dog said. "How could I be angry?" She told her how she had spent the rest of her short life looking for her. "I never found you," she said humbly, "but I never stopped looking." She told her a little about her search—just a little, with only the slightest allusion to camel dung. "You spoke to me every night," Noor said.

Then, remembering that Sharon had always had a hard time hearing what she told her, Noor spoke directly to Raphaela. "Tell her," she instructed, "that my time with her was wonderful, and that it elevated me. I had never known about love between people and dogs. Now I know what it is. That was her gift to me, and I will never forget it."

A Dog's Body

Raphaela spoke about another client whom she helped with the death of her dog. Shive was a pointer mix, 17 years old. She was stiff and in pain, no longer able to enjoy her life. Shive's family wanted Raphaela to ask her if she wanted euthanasia, and Shive said yes, she was ready.

"Shive, what messages do you have for your family?" Raphaela asked.

Shive said, "I'm old and feeble, but do you know, I still feel the same inside as when I was a youngster. I hate to go. It's been a wonderful life. Do you know why I came to you? It was to help raise Karl. Your son has turned out very well, and I am satisfied with the job.

"I came into the body of a dog because dogs are capable of so much love. It is a wonderful thing to be a dog, and this has been a lifetime of love, support, and service. I'm glad to be getting out of this suffering body, but I am still sorry to leave you.

"We all come back," Shive said. "You should know that the return isn't to the earth. When we're in our bodies we think that coming back means returning to the Earth, to our bodies. It isn't. The real return is to the spirit world."

She heard the same message from Gus, a 13-year-old dog with an interesting mix of Golden, Shepherd, Collie, and Chow ancestry. Gus had come to his person, Lorraine, as a stray when he was already an elderly dog in uncertain health. Lorraine consulted Raphaela to find out what she could about Gus's life before he came to her.

"I was scared," Gus said. "I have something wrong with my rear."

He didn't say so but Raphaela got the impression he may have been hit by a car. She told Lorraine, who said, "He does have a lot of trouble getting in and out of my car. Ask him how he feels now."

"Okay," Gus said. "I'm basically feeling all right, but I'm on a downward course and I won't recover. I'll stay around as long as I can. I love Lorraine, and I know she'll suffer when I go, so I'll stick it out, but I can't last. My pain is tolerable right now, so we can say whatever goodbyes we need to say. Then if it gets bad again, she'll be able to let me go."

The next time Lorraine called Raphaela for a consultation, Gus was gone. The vet had come and freed him from his suffering body. Raphaela called and he came in immediately and said, "You forget when you're in physical form what a relief it is to be out of your body!" He was much happier. He sent Lorraine a big, sloppy kiss, which she confirmed was exactly what he had always given her when they had been physically together. He showed a picture of an arched bridge, and showed himself being met by his friends, angels, and spirit guides as he crossed over. He said, "Tell Lorraine I will wait for her."

Lorraine, in mourning, said, "I want Gus to know that I wish I had done more for him."

Gus accepted her guilt feelings but said, "Please tell her it's not necessary! She did everything for me I could have wanted. I miss her just as much as she misses me, but it's easier for me to bear, because when you are in spirit form you can feel even closer to the people you love."

Having contact with beautiful souls like Gus, Shive, and Noor is one of the blessings of animal communication. By speaking to us so freely from the spirit world, they show that the curtain that divides us from the dead is not absolutely impermeable. To them, at least, it is gossamer. They have sympathy for our pain in losing them, and they share it, but they keep their connection to us long after they have left their bodies. This is one of my red-letter thanksgivings to Raphaela and the animals.

T Rex, the Service Dog

"Nothing inspires more reverence than the noble Service Dog. We see a sober, responsible seeing-eye Shepherd and we think, how noble, how good, how inspiring. Did I ever tell you about T Rex?" Raphaela said, settling down to her cup of tea.

"Was T Rex a service dog?"

"T Rex was a six-pound toy poodle. And yes, he was a service dog of sorts. He'd had a little bit of training as a hearing dog, but—well, I'll tell you about him. He was an interesting little character.

"T Rex's person was a woman named Jan. She had had a stroke and was now an invalid, confined to a wheelchair. T Rex was her constant companion. He was a little bundle of cheerful company, and she depended on him for that. He also kept her anchored in her personal reality. She only recognized her husband and her daughter intermittently, but she always knew T Rex."

"What kind of service did he do?"

"That was the problem. He'd misunderstood his role in her life, and thought he was supposed to defend her against all comers. He'd growl at absolutely anyone who came near Jan. Her husband could handle the dog, but her health care workers were having a terrible time. T Rex sat with Jan on her lap approximately twenty-three hours of every day. He got no exercise, and you know what a problem that is for all dogs, big or small.

"I was called in by Jan's daughter. Jan and her husband were spending several months with the daughter, and the daughter recognized immediately that T Rex was out of control. She asked me to speak with him, and I did.

"The first thing T Rex said to me was, 'I'm the privileged one!' He was very well named—a little tyrant king! He understood perfectly that he had a job to do, and that his job was to help Jan, but he had overinterpreted his mission terribly, and his ego had blown up to huge proportions

"I tried to explain to him that his actions were interfering with Jan's care. He said that Jan liked it, and I think that was true. She subtly encouraged the dog, even when he growled at her helpers. T Rex was expressing her anger and frustration at what had happened to her.

"I sent them an absolute garden of flower remedies: snapdragon and vine, which you know about, and also holly, which is for jealousy or the fear that others are getting more love and attention that you are. Also chicory, for selfishness and self-preoccupation. Chicory is a wonderful essence that encourages the development of 'selfless love, given freely, respecting others.'

"I also got Jan's daughter to take T Rex out for a walk every morning. You know how I feel about exercise, and I also thought it would do T Rex a world of good to have a little break from Jan each day.

"The flowers helped and the walk helped, and T Rex's service mania abated a little bit. Unfortunately, Jan and her husband went back to their own place and backslid. Then one day, about two months later, T Rex went running out their front door and was killed by a car. It was quite sad. They asked me to talk to his spirit, and he said, 'I was trying to go on with the walks.' He'd enjoyed his walks with Jan's daughter, and when they'd gone back home he thought he'd continue them on his own. It was a good thought, but he had no experience and no judgment—he'd never been out by himself in his life.

"I talked to T Rex, and he said he felt lost without his family. His last words to me were, 'You know what? I think I'll come back as their cat.'"

Trevor Steals the Show

I remembered overhearing Raphaela talking to a client whose dog, Trevor, was having some trouble with obedience. I asked her how Trevor was doing—had he gotten through his trials yet?

"Oh, Trevor," said Raphaela, bursting into laughter. "Didn't I tell you what happened?"

"No."

"Well, you remember Trevor was a Golden Retriever from Illinois. When I heard from his person, Linda, she was absolutely baffled. Trevor was an absolutely brilliant dog and he'd easily passed his CD. Linda was now trying to get him through Utility, which is a high level of obedience. He'd taken the test many times, and he always failed.

"What was so puzzling was that he always messed up on a different phase of the test. He'd do all the rest of it perfectly. One time he would fail the broad jump. Another time he'd ace the broad jump but retrieve the wrong article or go over a jump the wrong way. Once on the long down, Trevor rolled over onto his *back* and put his feet up in the air. It brought down the house, people were laughing so hard.

"By the time I talked to him, Trevor was famous. He would walk through the hall with his person, and the other competitors would call out, 'It's Trevor!' They were coming from all over to see which obstacle Trevor would blow this time. His person was getting a little desperate. She knew how smart Trevor was, and just couldn't figure out how he could flunk first one thing, then another. She had younger dogs that she wanted to bring to the trials, and here she was tied up every weekend with the one and only Trevor."

"So, what was going on?"

"It turned out that Trevor was a total comedian. He absolutely loved performing. He had figured out that once he finished his Utility Dog title, he wouldn't be going to any more trials. He couldn't bear that, so he'd come up with this stunt of always doing something wrong. He thought he was extremely clever, and I have to admit it—he was."

"What did you say to him?"

"I appealed to his show-biz nature. I said, 'Trevor, can you image what a great show it will be when you finally pass? The crowd will go wild. And Linda will be so overjoyed.' That didn't quite do it, so I negotiated with him: I said if he'd go through the whole test right, I'd convince Linda to keep on taking him to shows, even if he wasn't competing. He could think of himself as a mentor, helping the other dogs, especially the ones with performance anxiety. That was one problem Trevor definitely did not have.

"He went for it, and Linda agreed too. The very next weekend Trevor got his first qualifying score, and it wasn't long until he had his Utility Dog. It all worked out great. Linda took Trevor with her to all the trials, and she'd parade Trevor through the hall. He'd accept the applause of his public in a very gracious manner.

"Now that Linda knew what an outgoing, people-oriented dog Trevor was, she got him some more training and he became a therapy dog. Can you imagine? I bet he was wonderful at helping people. I don't care what might be wrong with me, a dog like Trevor would have to make me feel better."

"I had no idea dogs could be so funny," I said. "Is it unusual?"

"Not really. A lot of dogs are clowns at heart, but not all are as talented as Trevor. I was talking with Mary Getten the other day and she told me about a dog she talked to on the radio. She was being interviewed about animal communication, and the host of the show asked her to talk to his dog as part of the interview.

"She agreed, but asked that the host have her ask his dog something he really wanted to know—not a trick question to see if she was really communicating. Mary didn't mind doing it, but she hoped the dog wouldn't be boring and ruin her interview. The host said, 'Ask him if he's mad at me for letting him go outside that time.' It seems the dog had dashed out the door and been hit by a car. The car barely touched him, he wasn't hurt, and it was no big deal, but the host felt awful about it."

"What did the dog say?"

"Mary told the dog his person felt guilty. The dog shot back in a Groucho Marx voice, 'Let him! It's been getting me liver for breakfast for years!'"

Chapter Three

My Kingdom for a Horse

Prince William: Mind Pictures

As the smattering of unenthusiastic applause died away, Paulette turned Prince William back towards the barn. She felt tired and discouraged, and she could tell that Willie felt the same way. Not that she really cared how he felt. This was all his fault.

Didn't Willie understand that this was her last year to compete as a junior? If she wanted to win, this was her last real chance. She was 17. Next year she would have to compete as an adult amateur, against stiff competition. She knew she wasn't good enough to do well.

This was supposed to be her year. And Willie was blowing it for her. As they made their way through the little clusters of people and horses standing about on the show grounds, Paulette thought back to the course she and Willie had just finished. She could still see in her mind's eye the rail of the last vertical, lying on the ground where Willie had knocked it off.

All around her was the bustle of the enormous Harrisburg, Pennsylvania, horse show. There were hundreds of horses in her immediate view, some being walked up and down, some standing with their riders, others lining up to compete.

Inside the barn, Paulette dismounted and looked Willie over carefully. He was a big Thoroughbred with a reddish bay coat and a little white star on his forehead. When she and Willie were at home together, she thought him beautiful. But sometimes, like today, when they were surrounded with so many sleek, expensive horses, animals trained and groomed to the very peak of perfection and obviously accustomed to the best of everything, she felt that both she and Willie were rather inadequate. She hated that feeling, and looked crossly at her horse.

Willie's stomach contracted with tension. He had been receiving Paulette's mind pictures steadily, and they were not encouraging. "I'm not a good horse," he realized. "I didn't do well. She is unhappy with me."

He tried to think where he had gone wrong. He had tried so hard to follow Paulette's mind pictures as they took the last jump. Some horses were naturals at jumping. He'd watched them closely, and saw that they knew exactly where to place their feet before they took off. He wasn't like that. He could jump, but he needed Paulette to help.

This time, as they had approached the vertical, she'd pulled back on his reins. He felt the pressure on his mouth. It confused him. Did she want him to go, or not go? He opened himself to Paulette's mind, but found no clear picture. The fence was getting closer. He jumped, but not cleanly. The rail fell.

Now Paulette's mind pictures were perfectly clear. He had disappointed her again.

He hung his head, remembering the pain he'd gone through getting to this place. How he hated the horse trailer! He had to stand for all those hours, rigid, often frightened, always smelling the truck's foul exhalations. In fact, almost everything about competing was difficult for him. Now he and Paulette, accompanied by Paulette's mother Sarah, had been on the show circuit for almost three months. He thought longingly of home, and wondered when he would see his own barn again.

This particular show seemed even more intimidating than usual. The horses all thought themselves better than he was. No one would talk to him. Paulette was tense. If he did well at one of the shows, she would relax for a while, at least until the next show. This time, he'd done badly, and she was more anxious than ever.

Then a beautiful picture came into his mind. A field. Quietness. Green grass everywhere. A few quiet horses to talk to and soothe him. He tried to escape into that dream. He remained in this remote mood as Paulette walked him back to his stall. It was growing dark. Then his attention was drawn by the sound of Paulette behind him, talking with Sarah. He heard his name. He listened.

"Well, I think we should call her," Sarah was saying. "What harm can it do? Don't you want to know why Willie keeps knocking down rails?"

"As if he could tell us," said Paulette scornfully. "You know I don't believe in animals talking. But if you want to waste your money on it, I guess it's up to you."

"I would need you to talk to Raphaela too. You're the one who knows Willie best."

"Oh, all right. Go ahead, but don't expect me to believe in it."

Willie hardly had time to wonder what was going to happen when he heard a voice he didn't know, speaking directly to him inside his mind. "I'm looking for a Thoroughbred bay, eight years old, with a white star on his forehead, named Willie. He's at the Harrisburg horse show."

Why, that's me! he thought, astonished. He responded with a cautious, "Yes? I'm here."

"This is Raphaela," said the strange voice. "I'm an animal communicator. Paulette and Sarah asked me to talk with you. May I?"

"All right," Willie said politely, feeling more surprised by the minute.

"How are you feeling, Willie?"

"Me?"

"Yes. Are you feeling unwell at all? Your people are concerned."

"I'm tired," he confessed. "I feel very pressured. This show is a very big deal, you know."

"I do know. I have spent a little time at horse shows, although not nearly as much as you have, Willie. Whenever I go to one I do feel a little intimidated."

"There's so many great horses here," Willie said. "I'm kind of out of my league."

"Oh, no, you're really not. You're a wonderful horse. Do you have any physical problems that are bothering you?"

She must be asking about my back legs, Willie thought. Perhaps someone told her they used to hurt. "No, my legs are okay. I'm just tired."

"Ask him what would help," Sarah put in.

"Did you hear the question, Willie? What would make you feel better?"

He thought, and it came to him very clearly—his dream of the green field. "I would like to be someplace quieter," he told her. "A quiet atmosphere. Green grass. I'd like to have Paulette walk me around gently by hand. I know I would feel so much better."

"Ask him what he thinks about me." That was her—Paulette. He could hardly believe she didn't know his every thought, as he knew hers. Out of his love for her, he spoke very simply, without tact or disguise. "Paulette is wonderful, and I love her, but she does give me problems sometimes. She can ride aggressively and well, but other times she clutches at me." He told Raphaela what had happened on the last jump—the hand pulling back on his mouth and her weight in the saddle, just when he was supposed to go. "I think she gets weak. I need her to help me by seeing my takeoff point for the jump. I can't see it very well—I need her to see it clearly and show it to me. And she doesn't always do that."

"What did he say?"

Raphaela repeated his words to Paulette. When Paulette spoke again, after a silence, he heard once more in her voice the sweetness and humility he remembered from the days when they had been together at home, before they had started going to all these shows.

"Mom, it's true," she said. "I don't quite understand how Raphaela found out about it, but that is just what happened. I clutched at the last jump. I didn't want you to know, or even to admit it to myself, so I decided it must be Willie's fault."

The unseen woman said quietly in his mind, "Thank you, Willie. I've told them what you said, and I think it will go better in the future. Now I'm going to tell them about mind pictures, so Paulette will know how to help you better in the ring. I bet you see a big difference!" She thanked him for talking to her, said goodbye, and was gone.

Willie thought that she must have done a pretty good job of explaining, because right after that, he got his wish. For the whole of the next three days he and Paulette stayed out of the competition. He knew it was still going on because he saw the mind-pictures of other horses who were still competing nearby. But he and Paulette went instead to a quiet field a short distance away, and spent the time walking together quietly while he grazed, just as he had asked.

As he relaxed and unwound, he could feel the same thing happening to Paulette. As she walked about with him she talked, telling him how much she wanted to do well as a Junior and how grateful she was to him for taking her so far. He tried responding to her with his mind, and at times he almost felt that she understood. Certainly she felt his love for her. He was quite sure of that.

At the end of the three days, he and Paulette returned to the show. He rested quietly that night, feeling the excitement of the other horses swirling around him, no longer feeling ignored or scorned. The next morning Paulette came to his stall and stood with him for a while, talking to him quietly. She told him that they were entered in the Junior Jumper Stakes, and that it was a big course which would be a strenuous test of nerve and muscle for both of them. There were tight corners, several switchbacks and every sort of obstacle. "Think of all the obstacles you've ever seen, Willie," she told him. "This course has them all."

Willie was overjoyed to be taken into Paulette's confidence. When she swung up onto his back, he moved out into the warm-up area at a trot, feeling strong and happy. He concentrated on Paulette's cues and felt the energy and pictures flowing between them. They turned toward the first fence, and he picked up the canter smoothly. The first obstacle was a simple oxer followed by a vertical. This was easy! He flew over it.

Picking up speed, he and Paulette moved nimbly through the in and out, then surged forward over the broad water jump. Paulette watched alertly, counting strides before the takeoff spot, and Willie pushed off just where she showed him he should. Then he galloped up a steep bank and down a long grassy slope to a solid brick wall. Over it they went, a perfect unity of horse and girl.

Racing against the clock, they drove for the finish line. There was another in and out, and then came the last vertical. This was the spot where he had knocked down the rail three days ago, and the unwanted picture loomed before his eyes. Oh, he didn't want to do that again! Would Paulette be able to help him? Then he saw something wonderful in her mind. It was an exact picture of himself, sailing over the gate exactly in the center, with his front legs neatly folded up under him. "Right!" he thought. "I can do that!"

He sailed over the last gate perfectly. It was his best vertical of the whole course, and he heard the applause burst out from the stands, loud and sincere. As he cantered out of the ring he felt Paulette's hands stroking his neck. In her mind he saw a wonderful picture.

Himself. Prince William, the good horse.

Good Turnout

One morning in late fall Raphaela and I drove out to Pine Trails, the barn in Davis where she rides. We were going to talk with some of the horses, and I was to have a riding lesson. As we got into the car Raphaela glowed with excitement and pleasure. "This is probably my favorite thing to do," she confided. "Being with horses is every bit as thrilling to me today as it was when I first fell in love with them at the age of ten. One of the best parts for me of being an animal communicator is getting to talk to so many horses. And I love riding so much. Getting on a horse always makes me feel like a combination of cowgirl and princess."

I smiled at the picture. I was excited too, but also apprehensive. Horses invoke a complicated reaction in me, compounded feelings of longing and envy and frustration. My family always had dogs, and we once had a black cat called Dicy, which was short for Dicyandiamid—my father was a chemist and that was his idea of a joke. But horses were right out of our range. I remembered reading *National Velvet* as a young girl, and in high school I'd sometimes overhear a certain group of kids discussing their Saturdays at the Hunt Club. That was the romance of the horse, but the reality was different. The reality was Tony the Pony.

When I was six or so my family lived for a time in Texas. Tony lived in a field next door, and we were allowed to ride him, except Tony didn't exactly give rides. My younger brother and I would take turns climbing up on Tony's back. We'd sit there while Tony chomped grass. Eventually Tony, from boredom or aggravation, would buck us off.

Tony was just an extreme example of my riding experiences. To the extent that they could, all the horses I'd ever been on, mostly rental horses in various city parks, had behaved exactly the same way as Tony. *National Velvet* it wasn't. Recalcitrance on the horse's part, incompetence on mine—that about summed it up. I had long had dreams of a white-horse-on-the-beach, wind-in-my-hair experience, but so far it eluded me.

I knew Raphaela was a real horse lover, and today I was looking forward to getting to know horses as I'd imagined they might be. Raphaela had mentioned that she used to have a horse of her own, and as we zipped along the country roads I asked her about that horse.

"Actually I've had two. My first horse was an Anglo-Arab called Surym. He was a love. After he got too old and lame to be ridden, I gave him to the wonderful people at the vet school here at the University of California at Davis. They keep a riding school, and Surym was very happy with them. After Surym, I got Matrix." As she said Matrix's name she gave a deep, wistful sigh. I glanced at her and caught a look of profound regret on her face. "Poor, dear Matrix. I did so many things wrong with her." She sighed again.

This was my first intimation that having a horse for a companion is not all a canter on the beach. Indeed, it can be deeply problematic for the person and the horse. There are enormous rewards, but you make sacrifices to be together—

the horse perhaps sacrifices more, but the person has difficulties to endure as well. That's why people who have horses almost always have large animal families—dogs, cats, birds, turtles, rabbits. Every other animal is a piece of cake to live with, compared to a horse.

I asked if she would tell me about Matrix.

"Not now," she said—the first of many evasions. "Let's just say it's not easy to find the perfect situation for a horse."

"Why is that?"

"The horses say it best," she said. "They know exactly what they need. I'd say Deeter put it particularly well. Deeter is an absolutely magnificent Westphalian, about seventeen hands. That's big. He's a glamour horse, glowing chestnut with a white blaze and four white socks, which are very flashy markings. Linda, his person, does dressage and competes. Those white socks of his just twinkle in the ring, always very attractive to the judge.

"Linda consulted me to help her animals when she moved. Her household consists of herself, her two dogs, and Deeter, and they have just relocated from a large ranch in Grass Valley to a new house in Atherton. The dogs had adjusted well to the move, and Linda had placed Deeter in a lovely barn, the West Charter Oaks Barn, quite near her new home.

"The part of Atherton where Linda now lives is serious horse country. It's a wealthy neighborhood with lots of riding trails and several horse associations. Even so, it wasn't quite what Deeter had been used to. In Grass Valley they had lived on their own property and Deeter had been turned out on acres of pasture. Now, it wasn't like he was suddenly transported to the horse equivalent of a slum, or anything, but it was definitely a step down to be living in someone else's barn.

"Linda told me that shortly after he moved to West Charter Oaks, Deeter had gone lame. The vet had already been to see him a couple of times, but hadn't been able to help him much. Linda asked me to talk to him and find out what was going on. She suspected, rightly as it turned out, that his lameness had something to do with the move.

"Deeter came in telepathically, and I immediately sensed a very alert, intelligent, and outgoing horse. Not all warmbloods are like that, to be honest. I've talked to a few with a kind of aristocratic outlook, almost like royalty, that sometimes makes them a little snobbish about engaging in conversation with mere mortals like me. Luckily, Deeter wasn't like that, and we had a good talk.

"He started right in telling me what he'd been going through with the vet. When a horse goes lame for no obvious reason, the vet usually goes through a series of anesthesia blocks. They numb the lowest part of the horse's leg with local anesthesia. If the lameness disappears, you know the problem is there. If he's still lame, the vet then tries numbing the next section of the leg. You gradually work your way up the horse's leg.

"It's not a very brilliant approach, and it's tiresome and painful for the horse. The vet had been working with Deeter but hadn't succeeded in finding out where his problem was located."

"This is where you really need telepathic communication," I commented.

"Yes, well, that was the idea. After Deeter finished complaining about his treatment he told me what had happened. He was feeling extremely pent up and confined, and missed being outside. To relieve his tension he decided he needed to roll over in his stall. That's not a good idea, and he knew it. When a big horsc rolls over in a 12' × 12' stall, damage can be done. Do you happen to know how a horse stands up?"

"No, now that you mention it, I don't."

"He rolls into sitting position on his haunches, kind of like a dog sitting up. Then he uses his front legs as props to raise his rear quarters, finally straightening his back legs last. When Deeter was straightening out his front legs from the roll-over he caught his front hoof under a ledge on the edge of his stall. In his violent struggle to free himself he pulled his whole shoulder out of alignment.

"So that was the source of Deeter's problem. Linda verified it by looking in the stall. She and the vet found the exact spot where Deeter had caught his hoof. The wooden lip was gouged and the wood was chewed up.

"But now comes the part I found interesting. Deeter told me, 'In order to heal, I need to go outside. Staying in my stall all day blocks healing because it stops movement.'

"Deeter put his finger on the key to horse happiness. It's being outside and having the freedom to move."

I had a mental picture of a herd of horses galloping across an open field. "They really like moving around that much?" I said.

"It's not just that they like it. They *have* to do it. Horses are originally from the plains. Their bodies are designed for constant, leisurely movement. Wild horses spend around eighteen hours a day moving slowly over the grassland, nibbling on grass. They eat constantly, but only take in small amounts of grass—not a particularly rich food—at any one time. So *everything* about a horse, from his digestion to his circulatory system, depends on his being able to move and nibble constantly."

I remembered that when I was working with Raphaela and Mary Getten on the orca interviews, Raphaela had compared the plight of captive orcas to that of horses. Raphaela told me then that for an orca to be confined to a small pool is very similar to a horse in a small stall. Both of them suffer terribly when they can't move about freely.

"Do you know the expression *stocked up*?" Raphaela asked.

I shook my head.

"It means having swollen legs. The veins in a horse's legs depend on the massage action of walking to pump blood back to the heart. If the horse is standing still in his stall, blood just pools in his legs, and their legs swell. It's a problem stalled horses have constantly.

"Their digestion is another example. A horse in the wild would be walking, eating, digesting, and, of course, pooping, more or less all the time. Grass is mostly water and fiber, and it just moves on through him. Horses have no mechanism for vomiting, because normally they wouldn't need to vomit. Any food that didn't agree with them would soon be out of their system. Sorry to be so graphic."

"That's okay."

"So a horse that's kept in a stall just doesn't function right. He can't just suddenly adapt to such a completely different life. Instead of walking around for eighteen hours a day, he's lucky if he walks for two. For Deeter, a good day is one where Linda takes him out for a two-hour ride. The other twenty-two hours, he's just in the stall. And that's a good day! If Linda can't get out to the barn, Deeter's in his stall for a full twenty-four hours. He's fed twice a day on alfalfa, a rich legume hay, and oats, which are far more calorie-dense than grass. It totally overloads his digestive system. And psychological problems compound the situation, too. Deeter's problem was really psychological. He rolled over and hurt himself because he felt so confined in the stall.

"Almost every horse I talk to with physical problems puts in a pitch for more pasture time. Just the other day I talked to a horse named Marvin. He had such severe colic that he'd already had surgery *twice*.

"Marvin is a Thoroughbred who does the hunter circuit. When his person, Samara, called me, they were down in Florida for the winter shows. I should tell you that having a fancy horse like Marvin is a big deal financially. Samara had a lot of money invested in him, so of course she had to have insurance to protect her investment. The insurance company insisted Marvin have high-tech medical treatment, and that's why he'd had the surgeries. Abdominal surgery on a horse is an extremely big deal. Can you picture it?"

I tried, but the picture of a horse on an operating table overloaded my imagination.

"It's extremely difficult. Horses are so large that they can't lie down safely for very long. They start to crush their own lungs with their weight if they lie down for a prolonged period. In fact, they are built to sleep standing up, which is helpful to them for quick getaways in the wild. If you ever see a horse lying down, watch him for a few minutes. Usually you'll see him roll over a few times and then get back up. It's very unusual for a horse to stay down for more than an hour at the most.

"Any horse in surgery is at risk just from lying down. Then after the surgery he has to be immobilized, which just makes everything worse. I was very surprised when I heard that Marvin had been put through it twice. It just showed me that Samara and the vets considered his colic life-threatening.

"In the surgeries, the veterinarians found cyanotic intestinal tissues which they removed. The intestine was sewn back together, but when Marvin recovered from anesthesia, because of the shock and the handling, his gut was inoperative. Now, at the point when I was called in, Marvin had absolutely no gut sounds.

That means he wasn't digesting at all. He wasn't eating and he wasn't pooping. His system was totally hung up. He was losing weight and was very depressed.

"I tuned in to Marvin. He said he felt awful and had a stabbing pain in his stomach. That was obviously his incision. I figured they had already tried everything Western veterinary medicine had to offer, so, I thought, let's try something else. I'm a great fan of arnica, a healing herb that is used a great deal in Europe, so I suggested that Samara try giving Marvin arnica drops and rubbing arnica cream on his incision.

"I also suggested they try a program of bodywork called TEAM, which stands for Tellington Equine Awareness Method. TEAM is a program of bodywork developed by a Canadian woman named Linda Tellington-Jones. She grew up with horses and did all kinds of riding—hunter-jumper, dressage, hundred-mile endurance races, you name it. She studied the Feldenkreis system of bodywork for humans and adapted it for horses.

"I explained to Samara on the phone how to do a simple TEAM program on Marvin's ears. Traditional Chinese medicine teaches that the ears have points that reflect every area in the body. By gently working the ears, you can affect the horse's entire body. I suggested that after the ear work she might take Marvin out and walk him gently around for twenty minutes or so, then do a few more minutes of TEAM on his ears, then walk him some more. Many clients of mine have completely resolved their horses' colic with this method of walking and doing TEAM work. I also sent Marvin a long-distance Reiki treatment.

"The very next day, Samara called back. She said it was like a miracle. She took Marvin outside and did a session of TEAM, and Marvin immediately got gut sounds. Then she gave him some arnica drops, and he immediately pooped. She was overjoyed.

"When I tuned in to Marvin he told me, 'I have gut sounds!' He sounded so proud of himself. 'I have gut sounds!' Imagine that from a horse. It was so darling.

"Then Marvin said almost exactly the same thing Deeter said. He said that to get better he really needed to walk around. He also felt he should eat some fresh green grass. Well, grass does have a mild laxative effect, and I thought it might be good for him. We asked the vet and he gave Samara the go-ahead to try Marvin's program.

"Before I said goodbye to Marvin I told him that Samara wanted to know what he thought had helped him the most. He said, 'Tell Samara it's mostly her love. I loved it when she touched my ears. And the Reiki felt like a warm blanket.' He really was cured by a combination of movement, and Samara's loving, natural treatments."

We were almost at Pine Trails by this time. As we turned in the driveway I said, "What about Deeter? Did he get to go outside like he wanted, so he could heal?"

"Unfortunately, no. Linda felt it just wasn't possible."

I made an indignant sound.

Raphaela turned off the car engine and turned in the seat to look at me. "It's easy to criticize," she said, "but I've been there and I understand Linda's point of view. She loves Deeter very much and only wants the best for him. Compared to a lot of horses, Deeter's sitting pretty. But she felt the way a lot of people do—that she just couldn't keep her horse in pasture.

"If Deeter was in pasture, every time Linda wanted to ride she would have to slog through the mud or whatever, find him, clean him up, and bring him in to tack him up. She only had limited time for Deeter and that wasn't how she wanted to spend it. Also, of course, she was paying for a fancy stall. She thought Deeter was better off staying in it rather than roaming around the pasture, rolling in the mud, getting filthy and maybe trading kicks and bites with the other horses. That was her decision and part of me sympathizes with her. Unfortunately, Deeter and many horses like him are paying the price. And I hear he's still having problems. He's not lame anymore, but now he's started acting up. Linda called me again just recently about his bucking and biting. If he were in pasture, I know he'd be a lot calmer and easier to get along with. Basically, he has way too much energy and no way to burn it off.

"Deeter knows it, too. The last time we talked, I asked him about his biting, and he said 'It would be better if I had more turnout time.' I had to tell him it just wasn't in the cards. That's the reality of life. He might as well know it."

"Do you think working horses are better off? Police horses, cowboy horses, if they even exist any more?"

"That's a very difficult question. When is the last time you read *Black Beauty?*"

"I don't know. Forty years ago? I don't remember much about it except that I cried."

"You'd cry again, believe me. *Black Beauty* is the story of a lovely, well-brought-up horse. His kind family moves, and he is sold. He is then put to all the difficult, back-breaking jobs that horses did in those days. His looks and strength deteriorate, and he is treated with ever-increasing indifference and cruelty. There's a happy ending—Beauty is rescued and taken to live with kind people again—but the point is made. Anna Sewall, the author, was an early supporter of humane treatment for animals and her book was very much a morality tale about the mistreatment of horses."

"Deeter's better off, then," I said.

"Of course he is," Raphaela said thoughtfully, "but in the old days, when horses had to work so hard, they wouldn't have had the energy for the kind of nonsense Deeter's going through. They were just too tired. It's like our kids. Of course they're better off not having to go to work in the fields when they're ten. But lying around the bedroom playing video games has its downside too."

"What's the answer, then? Should people not have horses?"

"Oh, I'd never say that. But anyone thinking of getting a horse should definitely think very carefully about where the horse will live. Remember I said that ninety-five percent of my clients' dog problems would be solved if they had enough exercise? It's even more true for horses. If every horse had a quarter or half an acre of his own to run around in, they would all be so much healthier and happier. I bet my clients would save in vet bills whatever it cost them in real estate."

"Yes, and consultation bills, too. When you had Matrix, was she in pasture?"

"Some of the time. She's out in a large paddock now."

"Really? Where is she?"

"It's not easy for me to talk about her," Raphaela said. "Let's go see some horses." She shook her head as if to clear it, and her glow of pleasure returned. I followed her into the barn, more intrigued than ever.

Yoda: A Gift for Barrels

Yoda stood in his stall listening to Ashley talk excitedly to her friend Tiffany. He was too depressed to pay close attention, but when he heard her say "barrel racing" he felt a flicker of interest. Barrels were his specialty, and he remembered every detail of his training.

In his mind Yoda pictured the course. There were always three barrels laid out in front of him. His job was to run as fast as ever he could towards the first barrel. Then he had to slow down, but not too much, and make a circle around the barrel. Then he would race to the second barrel, and then the third, circling each one in turn. He knew how to stay tight in to the barrel, but not so tight he might slip and fall. It was a special talent, and he used to think he had it. Then he'd run like crazy for the finish line. Then there'd be the excited shouting and the applause, and he would be hugged and petted and made much of.

Oh, but that was before he'd come here. Before Ashley.

"Barrel racing is cool," Ashley was saying. "I could get into it, if only I had a decent horse."

"What's wrong with Yoda?" he heard Tiffany ask. "Isn't he a quarter horse? I've heard they're the best barrel horses."

"I guess. He's just not that pretty. I don't like chestnuts. I wish he was darker. And he's so short. Don't you think he's squatty looking? I wish he was more like Sandy, or maybe like Jeanette."

Tiffany sounded shocked. "Sandy and Jeanette are nice horses, but Yoda's yours! Didn't your mom just get him for you?"

"Yeah, and that's another thing. He's supposed to be trained for barrels, and I think Mom was gypped. I don't think he has the least idea what he's doing. He doesn't circle very well, and he's sooo slooooow."

"What about you, Ashley?" Yoda thought, anger momentarily obliterating his despair. "You think you're so great. You can't ride. You're too heavy, and you don't know how to balance on me, and you saw at my mouth, and, and . . . " Yoda's anger faded and he sank back into depression, with one final thought: "I'm all you deserve. And I'm not even going to try to be better."

"Yoda, talk to me. Yoda, are you there?"

If Ashley had been paying attention to Yoda she would have noticed that his ears flicked towards her, a sure sign that Yoda was listening. "Yes!" he cried. "Right here!" He was thrilled to be addressed directly, and it certainly beat standing here listening to Ashley complain about him, but he had no idea where the voice was coming from.

Then he heard Ashley's mother, Maria. She seemed to be speaking inside his head. "Do you have him?" she said.

"I do. Yoda, listen. My name is Raphaela. I'm an animal communicator. Your people have requested that I talk to you and ask you a few questions. Would that be okay?"

"All right."

"How are you feeling?"

"Okay. I've been better."

"Do you want to tell me what's going on? Maria's worried. She says you aren't working the way she knows you could."

"Is that what she thinks? You'd better tell her I can't do any better. I'm slow, and I'm bad at circling. And I'm not pretty."

"Who in the world told you that?"

"Ashley. She says so all the time."

Yoda heard Marie gasp. "I knew it!" she said. "Ashley's always saying awful things about Yoda. I knew he understood!"

The woman named Raphaela said, "I'm getting that his self-esteem is totally shot. He needs a lot of reassurance. Do you think you can do it?"

"Of course, I'd be glad to. It would be totally sincere, because Yoda really is a fabulous horse, but don't you think it should come from Ashley? He's hers, after all."

"You're right. I'll talk to her."

The next day Ashley came into his stall. Instead of ignoring him, as she usually did, she came right over to him and stood looking at him, right into his face. He breathed and waited. "Yoda," she said. He waited some more, looking at her attentively and carefully. Then he saw comprehension come into her eyes. "Yoda, I think you're really there," she said.

He didn't quite follow, but she went on, "Are you, Yoda? Do you really understand what I'm saying? Because if you do, I've been very wrong about you, and I'm so sorry. You are a beautiful horse. You're wonderful. I'm so lucky to have you, and if you'll only put up with me, I bet we can do something great together!"

With that she jumped up, raced through his grooming, and tacked up. In a few moments they were heading outside to the practice ring. He almost leaped with joy. She liked him! He raced for the first barrel. This was more like it!

Did Ashley still pull on his mouth? Sure! Was she still unbalanced? She was! But Yoda didn't care any more. She was a feather on his back—and if she wasn't, she would improve. He would do anything for her. He would make her a Junior Rodeo barrel racing champion! He would take her all the way to the finals at the big Rodeo in Galveston! And he did!

Pine Trails

As soon as Raphaela and I stepped inside the barn at Pine Trails we spotted Michelle Haseltine, the owner, deep in conversation with her newest horse. He was an aristocratic, long-necked Thoroughbred whom Michelle had just purchased at auction. "I only went to the auction to help a friend find a horse for herself," she said as she stroked the horse's beautiful neck. "My husband said to me, 'You're not going to buy another horse, are you?' and I said no, certainly not. But I came back with this guy. Isn't he gorgeous?"

He was. "I had to buy him," Michelle went on. "He was going to be destroyed because of back problems—you can see the sway in his back. I've been doing belly lifts for his back and he's already better."

I soon saw that Michelle has a huge heart for horses. It showed in every horse in her barn. Raphaela told me they were all extremely happy. They loved Michelle, who knew everything about them from their family history to the least particular of their movement. They all had great turnout, and plenty of room to roam around in.

The barn itself was nice but far from fancy—the money had obviously been put into land and into taking good care of the horses. It felt great there. This was the right place to begin my real riding career.

I followed Raphaela out to one of the pastures to look for Spotty, the horse Michelle had selected for me to ride. "Isn't she a charming person?" Raphaela said when we'd located her, standing around with several other horses. She was indeed. She was small, round, and placid, and gave me a quick flashback to Tony the Pony. Spotty might be small for a horse, but she was still big, strong, and formidable. We eyed each other, with what I hoped was good will. Then Spotty gave me a quick lesson in what active telepathic communicators horses are. I said, "Hello, Miss Spotty." I wouldn't normally be capable of spontaneously hearing a reply from an animal, but I clearly heard her say, "Make that Mrs. Spotty, if you please."

I relayed this to Raphaela, who said, "Oh, has she been bred? I didn't know. Let's check with Michelle." Michelle told us that Spotty had indeed given birth to a single colt, so I was careful to call her Mrs. Spotty from then on.

Madame Spot and I walked together to a small outdoor ring. I heard her say, in what seemed like a resigned tone, "I do this all the time." She meant, I think, that she is often chosen to introduce beginning riders to the basics of riding; I'm not sure if she was talking to me, or just noting the point to herself. It was interesting to be plodding along next to a talking horse.

The riding lesson went by quickly, except at the end, when it became painful. Concealing the state of my rear end as best I could, I listened carefully as Michelle explained that communication between the person and the horse is the very essence of riding. "Notice how Spotty's ears flick back towards you?" Michelle asked. "She's paying attention to you, trying to find out what it is you want her to do. You need to be careful to give her clear signals. You don't want to direct her in one direction, and look somewhere else. She can tell where you're looking, and she'll think you want to go there. So look where you want to go—she'll follow your attention."

Horses size people up quickly, she went on. Spotty knew everything about me within ten minutes, it seemed. This was not exactly reassuring, given my ignorance, but Spotty was patient and forgiving. She obediently walked and trotted round and round the little ring until an hour was precisely over; then, with the keen timing of the hourly worker, she went to the gate and stood there. Quitting time had arrived.

Raphaela and I walked Spotty back to her pasture, passing another pasture filled with horses on the way. "How does Michelle decide which horse goes where?" I asked. "It's interesting," said Raphaela. "Spotty's in the mare's pasture. That one over there is the gelding's pasture, and all the geldings are there except for Quiro. Quiro lives in the mare's pasture because he needs to be beaten on regularly to keep him in line."

"Beaten on?" I was shocked.

"Yes, and Shasta can be counted on for that."

"Who's Shasta?"

"She's the horse in charge of the mare's pasture—the dominant mare. In fact, here she comes now. Someone's had her out for a ride, and is just bringing her back. Watch what happens when she's let back into the pasture."

I noticed now that the eight or nine horses in the mare's pasture were lined up along the fence rail at intervals of about ten feet. Each one had her own eating place—a large tractor tire which had been recently filled with hay. Now a young girl opened the gate and led a small bay horse into the pasture. "That's Shasta," said Raphaela. "Now watch." The girl released the halter from Shasta's head, and she immediately went to the first tire, the one closest to the gate. Did it matter to Shasta that there was already a horse eating there? Not in the slightest. This tire, obviously in the prime, prize position, was rightfully hers. The horse moved away from it at once, and Shasta began eating from it.

"Don't the other horses mind?" I asked.

"No, not exactly. Shasta as dominant mare is the absolute ruler of the pasture. What she says goes, and the other mares accept that. Of course, she can always be challenged for her position. I think one or two of the mares in this pasture may have tried challenging Shasta—Michelle told me she saw what looked like a bite mark on Shasta's back the other day. But so far, Shasta's still on top."

"Is this because they are domesticated?"

"No, no. They're acting exactly as they would in the wild. Wild horses live in herds with an extremely hierarchical structure. Every horse has his or her place. Did you notice when Shasta took the first feeding station, that all the other horses moved down one tire? That's their pecking order, so to speak. And by the way, the leader of a herd is *always* the dominant mare. She makes all the decisions for the herd: where to go, when to leave, everything."

"What, not the head stallion?"

"No way. Stallions only stay with the herd until they're a year or two years old. Then they're driven off, usually when their mother has a new baby. They may live on their own or they might go live in a bachelor herd. Stallions do defend the herd against dangerous predators, and of course they spend a lot of time challenging each other for the right to be 'the stallion' and have breeding rights to the mares."

"Sounds like horses have a big social life."

"That's exactly right," said Raphaela, "and it's one of their biggest needs. It's just as important to a horse to have other horses around as it is to be able to move around freely. Elizabeth Marshall Thomas says that a dog by itself, without other dogs, almost doesn't exist. I would say that the same is true of horses, and it's another reason why it's so hard on a horse to be isolated in a stall."

As I took the loose lead halter from little Spotty's neck, she pranced away, making a little circle in her joy and then moving off to join the other mares. As she danced away, I thought once more of Matrix. What, in this joyous world of horses and the people who love them, could have happened between her and Raphaela to bring that look of sadness and regret onto her face? I resolved to keep asking Raphaela about her. Eventually she'd have to give in.

Duke: The Way Horses Do It

My name is Duke, and I'm the stallion around here. I know what you're going to say—Whoa! I've got to be some horse, right? I am, but you should know that it wasn't always that easy for me. I'm cooking now, and Carole, my person, thinks I'm fantastic. But there was a time not so long ago when she thought she'd wasted her money on me.

First you have to understand how I got my position. Being the stallion is the highest honor any horse can have, because the stallion is the one who takes care of the mares and gives them babies. Usually you have to fight the other horses for that right, but it was handed to me because of my qualities.

According to Carole, I'm loaded up with qualities. I'm handsome, and a good mover, and I'm also a nice horse. I have an excellent temperament, that's what I hear Carole say about me. She says it all the time, and she wants me to pass my qualities on to other horses. I'm glad to do it. I've got colts all over the place, as a matter of fact.

The part you'll have a hard time believing is the way I got all my babies. Not the usual way! I'm embarrassed to tell you this, but I do it with something they call a "phantom mare." I don't even know why they call this thing a mare, because it's really a contraption, not a mare at all. If you're a person, you're probably used to these kinds of things, but to me it is very, very strange. At first Carole couldn't even get me to believe that this contraption can give a baby to a mare a long way away. In the end, I had to take it on faith.

Even after I understood what the phantom mare was for, I just couldn't do what I needed to do. I thought maybe Carole imagined that I didn't *notice* that the thing wasn't a mare. That blew my mind. I'd just stand there looking at the crazy thing. How could Carole think I would be that dumb? That's when Carole started thinking she'd made a big mistake in making me stallion of the herd. I could tell, and it didn't look good for me at that moment.

Then Carole got an excellent idea. She brought in Raphaela, and we had a talk. I don't know if you know Raphaela, but she's a person who knows how to speak to horses like me. She reminds me of a horse herself, so I wasn't as embarrassed talking to her as I would have normally been, but it was still pretty tough on me. Raphaela asked me if I had any ideas that would help me do what I had to do with the phantom mare. When I thought about it, I did have a couple.

I told Raphaela that first of all, I wanted Justice to be around. Justice is one of my favorite wives, and just looking at her gets me in the mood. No other mare does it to me the way Justice does, so I told Raphaela it had to be her. I don't even care if she's in heat or not.

Then, I asked for music. I like classical music—that's another one of my qualities. Then, I asked if I couldn't just have a little time to stand there, looking at Justice and listening to my music, and thinking. I'm not going to tell you what I think about. That's going too far, and I didn't even tell Raphaela. But I have my ways of getting in the mood. That's all I'm going to say about that!

Well, Carole agreed with my ideas, and now we have it down to a system. Here's how it works. Carole puts on the music. I come out of the barn and go up to the breeding shed where they keep the phantom mare. Justice stands beside her fence, right where I can see her. She likes to flirt with me a little bit, and I like that just fine. Then I go right up to the phantom mare and mount it, but I don't really do anything. This is all part of my plan, you see. Then I go back and have another nice look at Justice. Then I stand there and think my *own thoughts*. Maybe you can guess what they are, and maybe you can't, but don't even ask because I'm not going to tell them to you.

Then I go back to the phantom mare, and this time in my mind it's like she's a real mare, and I give her a baby. I know—it's a strange way to do it! But that's people for you. I'm glad I'm not a person. The way we horses do it is a lot better. But I guess you wouldn't know about that.

Premarin Horses

Premarin (the name comes from Pregnant Mare's Urine) is the name of a little red pill containing conjugated estrogen. Premarin is taken daily by nine million women—it is the most prescribed drug in the United States. Every speck of the estrogen in Premarin, and also in PremPro and Premphase and Prempack—all brand names for different ways of packaging Premarin—comes from the urine of a pregnant Premarin horse.

From this tiny pill come great wonders. Stronger bones, hearts, vaginas, and brains—these are gifts given to women from the Premarin horses.

I had already been taking Premarin for five years when Raphaela came into my life. That's nearly two thousand tablets. Five years later, when we began working on this book, I had taken almost two thousand more. A few weeks ago I swallowed my last one.

It began when I asked her about working horses—the farm horses and mail horses and cavalry horses and cow horses that have given so much to build our civilization. Historically, horses have always worked very hard for us. They were a form of wealth, and a person who owned many horses was counted among the rich.

Intelligent people treated their horses well, as it was in their economic interest to do. People who were indifferent or unthinking or cruel just used them up. Either way, the value of horses rested on their ability to work. When a horse was no longer capable of working, its fate depended entirely on his owner's disposition.

I wondered how many horses today are working animals. The job market has definitely shrunk. Horses no longer do farm labor. The mail goes by airplane. We ride in horseless carriages. Cow horses barely exist outside of rodeos. The day-to-day work of a cow horse didn't have much to do with flashy rodeo tricks anyway. Roping and tying calves, and driving herds to market were seasonal, occasional activities. Day in and day out, what cow horses did was ride endless miles of fence lines, with the cowboys whose job it was to maintain them. Today, this work is mostly done from motorcycles and pickup trucks.

There are still some jobs available, though. The American Horse Council says that of the 6.9 million horses in the United States in 1999, 1.2 million were "working" horses. This category includes, in their literature, "farm and ranch work, police work, rodeo, and polo." Seven hundred and twenty-five thousand are race horses, nearly two million are show horses. The largest category by far, nearly three million horses, is recreational. In a sense, a recreational horse does have a job—the job of waiting for their person and then taking him or her for a ride.

Then there are the Premarin horses. There are about 65,000 of them today, but they do not appear in the Horse Industry Statistics put out by the American Horse Council, because most of them live on ranches in Canada. There are a very small number of them living in North Dakota and Montana, states bordering Canada.

You would have to call the Premarin horses working animals. They work every day, supplying urine to the ranchers who own them, who in turn sell it to Wyeth-Ayerst, the giant drug company that makes Premarin.

When I told Raphaela I was taking Premarin, she was not judgmental, but she was very firm: she needed me to know where it came from. She handed me a photograph of a horse hooked up to the complicated machine that collected its urine. By this time I knew the price horses pay when they are confined to stalls. It is much worse for the Premarin horses. They can't even move about within their stalls, and they are never taken out and ridden, or given a moment's peace in a field of green grass.

And on top of everything else, they are pregnant. Pregnancy is what puts the estrogen in their urine, so they are kept pregnant as much as is physically possible. When they give birth their colts are taken away, and most of them are sold at auction to be slaughtered for meat.

So that is the dreadful story. Actually, I knew it, although I preferred not to look at it from the point of view of the horse, but only from the angle of my own needs. I took Premarin mainly for the sake of my bones, because osteoporosis is a visible disease which I had seen in my grandmother and then in my mother. When I swallowed my Premarin tablet each morning I felt I was making my body more like a horse, strong and vital enough for me to ride it into old age. I thought the horses were lending me their strength, and I was grateful to them.

When I discussed hormone replacement with my friends, several had asked me why I didn't take "natural" hormones—meaning phytoestrogens derived from soy and yams. To me, hormones from a horse *were* natural. I felt closer to a horse than to a soybean. Strangely enough, taking horse estrogen had a primitive rightness to me. Even though the pill was obviously a highly refined product of modern pharmacology, I somehow managed to picture myself as a woman of the plains, deriving natural medicine from my faithful steed. I was back to the romance of the horse. It was time for a reality check.

I made an appointment with my nurse practitioner and took in Raphaela's pamphlet with the photograph of the Premarin horse. "Yes, I've seen this kind of material before, from patients who are animal activists. I've seen other material saying the conditions for the horses are not so bad," she said. "You and I will never know for sure, unless we go see for ourselves." Nevertheless she was quite ready to switch me to a plant-based estrogen, and wrote out the prescription on the spot.

So now I am depending on the soybean instead of the horse. It's not as inspiring, I must say. But I thought about my nurse's idea of going to see the Premarin horses for myself. I would like to do that, but in the meantime I had another idea. I picked up the phone and called Raphaela. "I know we don't have a specific horse to talk to," I said, "but would you be able to contact a Premarin horse?"

"I can try," she said. I told her I would like to express my gratitude for all the help they had already given me, and learn for myself what it was like for them.

Raphaela put down the phone and quieted herself for telepathic communication. I tried to do the same, hoping I could hook on to her telepathic link and get some direct communication myself.

After some time Raphaela came back on the line and said, "I have someone who's calling himself the Spirit of the Horse. He seems like a masculine presence."

"Could you tell him that I want to thank horses for their power and strength?"

A minute later Raphaela burst out laughing. "He says, 'We can send her plenty of power.' I said, 'Oh, would you do that?'"

Another pause, and then Raphaela said, "I explained what you meant about the Premarin, and he's responding. He says that yes, it is a service that horses are performing. What they are giving is the essence of our mares' femininity. Why don't you get it from the vegetable world, the way horses do?

"He's saying something more. He says they are not totally against doing this service for us, but the conditions are too hard. It's so typical of humans, he says—we always go too far. We think, 'this is good, how can I get more?' We find a good plant, then we manipulate the environment so nothing grows but that plant.

"Now he's telling me that horses don't mind sharing their urine, but they are terribly uncomfortable. Their legs stock up. They are sitting ducks for colic, hoof problems and respiratory infections—it just goes on and on. Horses are used to hard work. They have always worked hard. This is difficult, but at least here there is plenty of food. I'm getting that it hasn't always been the case that they've gotten enough to eat."

"In that sense, they are well taken care of," I said.

"Yes, and of course it wouldn't make sense to treat them so badly they couldn't produce for you, would it? It's interesting that he doesn't condemn the whole Premarin operation completely. I'm going to ask if he can connect me with an individual Premarin mare so I can ask her about her conditions. Hold on.

"Okay, I've got someone. She says her name is Iris, and she's a gray Percheron, about six or seven years old. She's telling me, 'I love to eat. It's not so bad here. I eat all the time.' She's showing me that she has had other lives of confinement. She used to be in a bigger barn than this one she's in now. All the animals were just milling around in the barn, rather than being in individual stalls. There were bitter storms outside. She says, 'This is better—it's cozier here.'"

"Could you ask her what could be done to make her conditions better?" I asked Raphaela.

"I'll ask her, but I'm not entirely happy with that approach. Just because Iris chooses not to condemn what we're doing to her doesn't mean that it's right. Hold on. Okay, Iris has referred me back to the Spirit of the Horse, and he's having another laugh at my expense. I guess I was showing him a ridiculous picture. I was thinking that the horses would have some freedom to move around if they weren't tied to the machines, and I pictured each horse having a personal attendant to follow her around with a bucket to collect her urine. The Spirit of the Horse said, 'that would certainly solve your unemployment problem.' It seems there are something like seventy or eighty thousand Premarin horses. No, he says, there's only one way. 'Stick with the soybean.'"

As Raphaela talked, a picture came into my mind; I'd like to think it came from the Spirit of the Horse. I saw Premarin creating a spiritual link between women and horses. We are all part of an enormous circle, formed out of the horses who give this service and the women who accept their gift. The right thing to do would be to complete the circle by giving Premarin horses their freedom.

We have the absolute power to do this, if we all decide together. If no woman takes Premarin, it would no longer be made.

We have been given a gift of years full of strength and life. We can use our strength in any way that seems good to us. We have plenty to spare. Why shouldn't we use some of it on behalf of the horses who gave it to us?

As for me, I've taken almost four thousand Premarin tablets. I owe a very great deal to every Premarin horse. There's not much I can do to pay them back, but what little I can do, I will. Send them my gratitude; tell their story; and stick, from now on, with the soybean.

Matrix

After I'd asked her four or five more times, I finally convinced Raphaela to tell me about Matrix. Then I finally understood how complicated her feelings for that horse were.

Matrix was born on a small breeding ranch owned by amateur breeders who raised four or five colts every year. When Raphaela first met her, she was four years old, and had spent most of her time out in the pasture, doing as she pleased. She was "green broke"—you could sit on her, and she knew how to walk and trot and canter, but that was all the training she had had.

Raphaela says that if she ever gets another horse it is going to be a weanling. The best way to bring up a calm, unflappable horse is to take her about with you everywhere, like a puppy. She has a friend who even walked her colt right into a hardware store as a training exercise. This way the colt has good experiences of all the things that can otherwise frighten horses half to death: honking cars, jumping kids, flower leis, fire trucks, whatever.

Matrix didn't have the advantage of that kind of upbringing. Spending her formative years in a remote pasture was natural for her and nice at the time, but it didn't prepare her for life in the real world. She was spooked by everything. Blankets terrified her. The *same* blanket hanging on the arena rail terrified her over and over; she saw it 50 times without it ever becoming her friend. She had no idea how to act around the shoer. This is something a horse has to be taught at an early age, and Matrix hadn't learned it. She repeatedly jerked her hoof away from the shoer. Once she even reared and came down on the foot of the shoer's assistant, breaking three bones in the poor man's foot.

Another time she was tied to a support pillar in a barn. She spooked and pulled back so violently that the 25-foot pillar was nearly dislodged.

However, she was very, very beautiful—a deep bay color with a black mane and tail, and black feet. She had the loveliest movement Raphaela had ever seen, and a soft, beautiful eye. (She had worry lines above her eyes, but Raphaela found it possible to overlook those.) Raphaela fell in love with her at once, and decided to buy her. She took Matrix to Piedmont Stable, a beautiful old barn in Redwood Park, where she had been riding for the past 15 years.

Raphaela still shudders when she thinks about what happened next. Matrix was not used to living in a stall, and she didn't like it. She kicked at night and once kicked a panel out of her stall. There was splintered wood everywhere, and the horse in the stall next door was understandably appalled. She rolled in her stall when Lloyd, the barn manager, put in fresh shavings, and got cast regularly—stuck in a lying-down position too close to the wall to get up. Lloyd was at his wits' end with this. Matrix continually reared when tied in the cross ties, and once even went over backwards.

Nor was she easy to ride. Startled by everything (a bird landing 20 feet away would do it), she would buck, dance, kick, and bolt. It wasn't that she wanted to get rid of Raphaela, she simply could not control her fears. Raphaela broke numerous bones.

Looking for a solution, Raphaela moved Matrix to a barn in Martinez, where there was more turnout. This was an improvement, but still not ideal. The barn had too little pasture for too many horses, so the horses would be let out in different small groups each day. This created a new problem. Horses are intensely hierarchical within their herds. Each time a new group of horses went outside together, they would jockey and fight for position until a hierarchy was established. Then, the next day, a new group would go outside, and the whole process would have to start again. Matrix participated enthusiastically in the fighting; vet bills began to pile up.

Even when the social structure was set, Matrix was so happy to be let out, and had so much energy, that she would run around madly, colliding with fences and other horses. More vet bills. Raphaela again noticed the barn owner giving her that peculiar look that says, "Maybe your horse would be happier somewhere else."

Maybe so, but at Martinez Matrix did settle down a bit. Raphaela and Matrix worked regularly with an excellent trainer. She and Matrix began to make improvements together. Matrix turned out to be a smart, sensitive horse who learned quickly. Her movement was still totally spectacular. She and Raphaela began competing, but their progress was slow. Raphaela explained that the problem was, green horse, green rider. She had to be taught the movements before she could teach them to Matrix, which made their rate of progress excruciating. What Matrix needed most was a better rider.

Raphaela's regrets are partly that Matrix never fulfilled her potential, but mostly that she was so unhappy. Matrix learned to hate dressage, which she found intensely boring, and disliked trail riding, which frightened her. Raphaela feels Matrix would have loved dressage if she, Raphaela, hadn't learned so slowly that Matrix became bored. "I wish I had had the courage just to *give* her to one of my great trainers," Raphaela told me. "I shouldn't have even thought about the money, it would have been so much better for Matrix. I used to think I'd never sell an animal. But in the end, that's what I did to Matrix."

"Were you already an animal communicator back then?" I asked.

"I got into it right around that time," she said. "In fact, it was partly because I was having so many problems with Matrix that I was so driven to learn telepathic communication. I thought if Matrix and I could really talk, it might help me do better by her. I belonged at that time to the California Dressage Society, and they're the ones who sponsored that talk by Beatrice Lydecker where I first heard of animal communication. I couldn't make it to that talk, but just knowing that people were doing animal communication made me totally determined to learn how to do it.

"Shortly after that I connected with Penelope Smith and began working with her. Penelope and I both talked with Matrix a lot. I was also learning a lot about various kinds of bodywork—TEAM, flower essences, homeopathy. I did everything I could think of with Matrix, but the kind of problems we had together couldn't be solved with bodywork. She was just too much horse for me. She would have been a difficult horse for anyone, but for an amateur like me, she was just impossible.

"After I sold her, I kept in touch with her telepathically. Once I even ran into her physically in a barn in Pleasanton where I'd gone to consult with some horses. It was wonderful to see her, but totally heart-wrenching. She thought I had come to get her and bring her home. I cried. That's when I decided it would be better not to keep contacting her telepathically—it was too upsetting for both of us."

"Would you mind if I tried talking to her?"

"Do you think you could do it on your own?"

"I'd like to try. I seem to do better with horses—they're so receptive, and I seem to hear them pretty well too."

"All right, then, try it. Be sure to give her my love if you do reach her." She gave me the particulars of where Matrix was, so that I could reach her through telepathic channels.

Several weeks went by before I finally worked up the courage to do it. It was late at night, and I lay in bed and tried to remember everything Raphaela had taught me. First I quieted my mind with a few minutes of meditation. Then I said silently, "I'm calling a bay horse named Matrix. She lives at Air Dance Farm, in Pleasanton, CA. She's 17 years old and is 16.3."

The telepathic operators must be terrific, I thought. No number needed—and think how many horses there must be in the world.

"I'm Matrix," said a faint voice in my mind. "Who are you?"

"I'm a friend of Raphaela's," I said, thrilled to the core to be speaking with the famous Matrix at last.

"I know her!" said Matrix, very proudly. "Raphaela! Yes, I know her very well."

"She's told me all about you, too, Matrix. She asked me to give you her love. May I ask you some questions?"

"Go ahead!"

I wanted to ask her about the difficulties I'd heard about, but didn't know quite how to start. I shouldn't have worried. In telepathic communication you don't have to pick your words and phrase things just so. It's telepathy, after all—the animal hears what you're thinking whether you spell it out or not It can be a little unnerving, but at least it works. As my mind floated over what Raphaela had told me about her, trying to pick out the most diplomatic way to start, Matrix said, "You're thinking about all the things that happened. Yes, it was hard at times."

Yes, I thought, it certainly sounds like it was.

I expected to hear from her something like what I'd heard from Raphaela—her version of what had happened, perhaps regrets similar to Raphaela's, perhaps a sense of unrealized potential. I waited, readying my sympathy.

I heard nothing of the kind! Matrix's reply was strong, open, completely forthright and very loving. "Does Raphaela think we were together for some small reason?" she said. "Does she think I came into this life to learn those little steps?"

As Matrix said this she showed me a sublimely funny picture of a great, huge horse doing dainty little pirouette steps as she danced around a ring.

"I could do all that if I wanted. That wasn't what my life was for."

She paused for a minute as if collecting her thoughts, then said, "Raphaela and I each came forward from our herds to learn to speak with each other. This is something great, not something small. It wasn't easy for me. It wasn't easy for her. We had many layers to break through. Our herds have been divided for many, many years. If we had just stepped into the ring together, it would not have happened. We would have been caught up in all that. No, everything that happened was just right."

Another pause, then she said, "Raphaela should call me. Tell her, don't be afraid. It no longer upsets me to speak with her. I am resigned to our separation and I am very happy with my people here. Tell her I will help her. She should call me any time. Be sure to tell her this."

Then a wonderful thing happened. Matrix sent me an experience of her physical presence. It was as if she was really there, right in the room with me. It was the fulfillment of my dream: the romance and the reality of a horse, both together, realer than real. It was magnificent. I wasn't on the beach with the wind in my hair. I was in my own bed, but I was also with Matrix, and the air that flowed around us was the pure air of Paradise.

I was still marveling at her incredible beauty of body and soul when she faded from my mind and was gone.

Chapter Four

Leviathans

Gray Whales Listening

The article in the *Eureka Times-Standard* caught my eye immediately. "20,000 Gray Whales Missing." That sounded serious. Gray whales aren't especially easy to lose, I've always felt. I picked up the phone and dialed Raphaela.

"That's great news," she said when I'd read her the headline. "The plan may be working."

"What plan? Working how?"

"The whales may be avoiding Makah hunters. Oh, I hope so!"

Macaws? Gray whales are being hunted by birds, now? I demanded a fuller explanation.

"Let me just finish with this client," Raphaela told me, her voice filled with satisfaction. "I've got a person on the line who's trying to switch her cat over to freshly cooked organic chicken. The cat only wants to eat dry cat food. We're negotiating."

A few minutes later Raphaela came back on the line. "You know that commercial whaling has been banned by the International Whaling Commission since 1986, don't you?"

"Yes, of course."

"That's great, but the ban is nonbinding and the Commission has no authority to enforce it. Nations that want to whale just go ahead anyway. Japan and Norway are active whaling nations, but international public opinion is solidly against them. No one likes bad press, and Japan, especially, is always looking for ways to weaken anti-whaling sentiment.

"One way to do it is to try to attach the label of "traditional" to whaling practices. People feel that there's something more acceptable about whaling if they can think of it as part of an aboriginal hunting tradition."

"I know. It's easier to romanticize," I agreed. "So?"

"The Makah are Native Americans who live on the Washington coast, in a place called Neah Bay. They've actually been able to obtain a permit from our government to kill five gray whales per year for the next five years. The permit came through just in time for the grays' migration from Alaska to Baja California this year. They claim that whale hunting is a vital element in their tribal customs."

"Is it true?"

"No. The Makah did hunt whales in the past, but they haven't had a whale hunt in 70 years. The tradition is long gone, and the tribal elders know it perfectly well and are opposed to it. I've got a copy of a letter from seven Makah elders to the Whaling Commission. Wait, here it is. Let me read you part of it. 'The whale hunt and other important issues were never brought to the people for a vote, or simple notification. . . .We believe the hunt is only for the money.'

"The oldest Makah is a woman named Isabel Ides. She is the only one who remembers the last traditional whale hunt, 70 years ago. Here's what she wrote. 'Nobody knows how to eat it any more. Nobody knows how to hunt it any more. There is no need for this.'"

"Why are they doing it, in that case?" I asked.

"It's pretty certain that the Japanese whaling industry is behind it. They want to use the fact that we've granted a permit to the Makah to undermine our moral authority to oppose whaling by Japan. Then they'll turn around and say Japan also has a whaling tradition. Also, Iceland is watching closely. They'd like to resume whaling themselves, but so far we've been able to pressure them out of it. Now they'll be able to say that we permit it for the Makah, and whaling is just as traditional for them as it is to the Makah."

The thought of gray whales being hunted appalled me. I remembered a story from Michelle Gilders' *Reflections of a*

Whale Watcher that showed how thoroughly trusting the grays have become since people have been visiting their calving grounds in Baja California. Gilders herself actually stroked a baby gray's baleen, the screen in his mouth he uses to strain sea water, while the whale held his mouth open and all but purred. Hunting such a trusting animal struck me as unthinkable.

Raphaela hadn't sounded sad, though, and there was the tantalizing headline about the missing whales. I waited for her to go on.

"When the Makah permit was approved," she said, "there was outrage in our community. At first we couldn't think what we could do, but then Mary Getten came up with a great idea. She enlisted every animal communicator we know, and they all sent out a message to the grays. 'Stay away from the coast! Just stay as far out as you can! The coast is very dangerous.' She even sent the communicators a map, so they could visualize the dangerous area and send the message to the grays.

"When you read me the headline, I guessed the whales heard the message and are changing their plans accordingly."

"The whales haven't disappeared, then? Where are they?" I asked her.

"I don't know exactly. They may be taking some kind of evasive action."

I sent a prayer for the deliverance of the whales, and began watching for more news. I used the Internet to scan daily for reports on the gray's migration. A few days later I saw this article in the *Seattle Times*:

Gray whales delay migration;
warmer water may be cause

Gray whales that summer in the Bering Sea but should be headed to Mexico by now instead are massing off Kodiak

Island, leading scientists to speculate that warmer ocean temperatures may be responsible.

"It's very unusual that they're here so late in the year," said Kate Wynne, a University of Alaska marine mammal biologist. . . . Meanwhile, a Seattle biologist says she is waiting for the whales to arrive in force off the Washington coast.

"In surveys we've flown to 30 miles offshore we have seen very few gray whales," said Sue Moore of the National Marine Fisheries Service's Alaska Marine Science Center.

Reports from Oregon say few whales have passed by there, too. The Makah Indian tribe at Neah Bay has been waiting for the migration to resume a whaling tradition that has been dormant for seven decades.

I called Raphaela. "The grays are still in Alaska," I told her. "Does this mean the whales definitely heard the message?"

"I can't be positive, but it's looking more likely," she said, and asked me to keep checking for further reports.

The next day I spotted an article from *The Oregonian*.

Sightings Signal Gray Whales' Visit

An aerial survey off Oregon's coast spots a few of the creatures, and other reports indicate they're more plentiful off California.

"I guess I'm surprised at how many are offshore because the weather has been so nice near shore," expert Bruce Mate was quoted as saying. "Typically they go offshore more when we have storms and there's big swells."

"What if there are Makah waiting? Does it say anything about that?" Raphaela chortled on the phone. I continued reading:

In California, reports from Oregon that trained volunteers were not seeing gray whales as they usually do during Christmas week worried whale-watching enthusiasts. But it appears that some whales slipped by without being seen from Oregon.

"Go grays!" we chorused.

Gray whales are being spotted in healthy numbers off the California coast—including by Bodega Bay fishermen—indicating that their annual migration from the Bering Sea to Baja California is on course after all. . . . "We're in the peak of the southern migration," said Doreen Moser, assistant director of education at the Marine Mammal Center in San Rafael, California. "The fact that the Oregonians didn't spot them means maybe they were further offshore."

"That clinches it," Raphaela said. "The grays were definitely avoiding the danger area. They swung out wide to miss Neah Bay, and stayed wide for safety's sake as they came down past Oregon."

Finally, on December 30, 1998, I was able to read Raphaela an article posted online by the *Sea Shepherd Conservation Society.*

Gray whales: 5 . . . Makah Whalers: 0

The Makah Tribal whalers of Washington State failed to take a single gray whale from the 1998 quota allotted to them by the United States government. "This translates into five whales saved . . . these five whales cannot be added to the 1999 quota," said Captain Paul Watson, President of the Sea Shepherd Conservation Society.

Watson also revealed in this article that the conservationists on the *Sea Shepherd* had also tried to warn off the whales by sending out underwater sound transmissions of orca sounds from their ship. Orcas do lurk near the shore in hopes of catching a baby gray whale, so hearing their sounds might well deter the grays.

"I didn't know that," Raphaela said thoughtfully. "It sounds there was some good teamwork going on. First the grays got a telepathic broadcast from the animal communicators warning them about the Makah. I believe it must have reached them in Alaska and delayed their departure for the south. Then, as they swam along the coast, keeping farther than usual out to sea, they caught the underwater sound transmissions from the *Sea Shepherd* anchored at Neah Bay. Only when they knew they were safe did they come back to their usual path near shore.

"Everyone did something to help," Raphaela concluded. "It was brilliant of Mary to think of using all the animal communicators to warn the grays of the danger. The communicators did a great job, and the *Sea Shepherd* crew did too.

"I have to say, though, that the real credit goes to the grays themselves. They're the ones who made the decision. They could easily have ignored a warning coming from our species. I for one wouldn't have blamed them. Thank heaven they chose to heed the warnings. I am very moved by their trust in us. That's what saved them.

"At least for now," she added. "The Makah permit goes for four more years, and the grays pass by Neah Bay twice every year. We'll have to watch and see what happens when they go back north in the spring."

Humpbacks Aloft in the Caribbean

Early in the spring of 1999 Raphaela called with more whale news. She had decided to take a trip to the Dominican Republic to the breeding and calving grounds of the humpback whales.

The humpback whales she would be visiting spend their summers in the North Atlantic and swim to the Caribbean in the winter to mate and to give birth. This works out nicely, since a humpback pregnancy lasts just about a year, and many humpbacks do appear to have a calf every year.

The trip was organized by a woman named Sierra. Sierra is a devotee of marine mammals, particularly dolphins, who loves nothing more than sharing her knowledge of and affinity for all cetaceans with others. The trip was to be aboard a catamaran called the *Bottom Time 2,* one of only a handful of boats given a permit by the Dominican Republic to take passengers out to watch the humpbacks. There would be lots and lots of whales, including many babies. Sierra had promised that they would even be able to swim with the mother whales and their calves.

"It's too fabulous a trip to pass up," Raphaela said. After the time she had spent interviewing wild and captive orcas and the successful message to the grays, she was eager to meet their big brothers the humpbacks. She felt that the wisdom of these magnificent creatures, whose scientific name is *Megapteros*—great winged ones—and whom the orca elder Granny always called "winged whales," would help her communicate more deeply with all animals.

What made it even more enticing was the fact that no less than nine animal communicators, including five of Raphaela's own former students, were going. Whenever a group of animal communicators gathers, animals often share their wisdom more freely. An atmosphere of love and trust is created, and communications fly between the species. It's like a good party with lots of lively people in attendance. Even the shy wallflowers get drawn into the action.

Nine animal communicators? Thousands of whales? I agreed with Raphaela that it sounded like the perfect trip. Alas, everything about it was perfect but the timing. My work would keep me in California all spring. I asked Raphaela to keep her journal faithfully, so that I wouldn't miss any detail of her journey.

Raphaela's Journal: Sunday, March 21, 1999

I made it to the San Francisco airport yesterday morning on time (just!) and found my dear friends Jasmine, Bonnie, Carla, Mary, and Bay standing around together near the ticket counter. Formerly my students, many of them have become colleagues and great communicators in their own right. I had a shopping bag full of snorkeling equipment with me—my carry-on. "I don't even know what most of this stuff is," I told them as I peered into the bag. The truth is I was worried. I hadn't had time to try on the mask or the flippers. I just took what the nice man in the dive shop told me to buy.

I was miserably ill with the flu for three days before the start of the trip, and almost canceled. I don't get sick often, and when I do I always wonder whether the illness is telling me something. Was this extravagant trip a wild goose chase after something I could just as well get at home, among my horses and rabbits and birds?

I like my comfort, and Sierra hadn't been very reassuring on that point. She said the cabins on the *Bottom Time 2* had all the luxury you'd expect in a horse stall. She suggested we buy our own 50-gallon drum of fresh water, so we'd have something decent to drink. She reminded us that we'd need to know how to snorkel, which is not something I was born to do. What had I let myself in for?

When I had finished checking in and was waiting in the departure area, I noticed that I suddenly felt much, much better. That had to be a good omen. In fact, I think it was whale magic. "Whale magic" is a well-known phenomenon, and quite real. Whether you're a cool, objective scientist or an avid whale devotee, you can't help but recognize that whenever there's a whale anywhere in the vicinity, you enter a sort of Twilight Zone of strange coincidences, wonderful events, and strangely enhanced energy. Yes, my sudden return to health was whale magic. My initial enthusiasm for the trip returned, then doubled.

In Miami we changed to a small plane for the flight to Puerto Plata. I knew Miami was tropical, but we were hermetically sealed within the airport. It was only when we finally we stepped off the plane in Puerto Plata that I understood what the tropics were. The moist air greeted us like an embrace. The soft fragrance of hibiscus flowers enveloped us, our eyes rested on graceful palms, and the skin of our faces relaxed and softened in the moist air. Everyone suddenly looked five years younger. I breathed in the fragrant night with relief.

Our group, now 20 strong, was greeted by Pedro, the driver who was engaged to take us to the *Bottom Time 2.* We made our way through the tiny port and onto the ship, a beautiful three-decked catamaran. There we met Captain Roger Maier. We were shown to our cabins on the lowest deck. As I stepped into what would be my home for the next week, my heart sank. Sierra had been generous in comparing the cabin to a horse stall; at my barn the horses live in spacious luxury by comparison. And that shelf—surely I wasn't expected to sleep there? But if not there, where?

I was by now thoroughly exhausted. We'd had to wait for our luggage, and there had been no time to secure the water. I'd have to drink whatever water there was on board. I'd already been informed that the next day would be difficult: we would leave very early for the approximately six-hour trip over open ocean to the humpback breeding grounds. The *Bottom Time* was fast but the passage would be rough. We should expect to be sick. But I didn't care—I was still thrilled and excited. Somehow I already knew that the discomforts would be minor, and the rewards incalculable. "Whale magic," I thought to myself as I crawled onto my shelf and went to sleep.

The *Bottom Time* took off about 5 A.M. this morning, so we could all start getting seasick even before we woke up. Seasick I was. The weather was calm, but we were moving through the swells of the open ocean. One huge *whump* following another for what felt like days, although it was in reality only the promised six hours. I lay in my bunk feeling distinctly green. My dear friend Carla, with whom I was sharing the cabin, was up repeatedly visiting the washbasin. We alternated between groaning and laughing at our predicament.

Finally the *whumps* stopped, and I crawled off my shelf and made my way onto the now only gently rocking upper deck.

Much better! The *Bottom Time* dropped anchor, and one by one the other passengers appeared. I fell into conversation with Teresa Wagner, who is a dear friend and one of the two "official" animal communicators on the voyage's staff. As we were exchanging news, Sierra called the group together in front of two small boats resting on the deck.

Indicating the calm waters stretching in every direction, she said, "We're now about 90 miles out of Puerto Plata. This area is called the Silver Banks. It's a very extensive humpback calving and breeding area. You may have heard that humpbacks have made a great recovery in the Atlantic Ocean. It's true. There are thousands of whales here now. These whales spend the summer around Stellwagen Bank, in the northern Atlantic, and each winter they migrate down here.

"The water is full of reefs, and is very shallow in places—80 feet, 60 feet, 30, 20, even 10 feet deep. Some places are so shallow you can barely take a rowboat. There are lots of wrecked ships in the area. Before the area was thoroughly charted, striking a reef was a constant danger. Back in the sixteenth and seventeenth centuries, Spanish ships might run aground while carrying silver back from the New World. That's why it's called the Silver Banks."

She then pointed to one of the two small boats, a small inflatable raft. "This is the Rubber Duck," she said, "And this one is called the Knife." The raft, of course, instantly became the Rubber Ducky. The Knife was a long, slim boat, bigger than the Rubber Duck, with a canopy over the center to keep off the sun. "We'll be dividing up into two groups each day and going out to wherever the whales are. Each boat will have one of the official animal communicators in it." She introduced Teresa and the other official communicator, my former teacher Penelope Smith. "I know many of you are communicators as well, so I'm looking forward to some interesting conversations with the whales, " she said.

A beautiful, tanned blond woman, muscular, compact, and radiating competence, then stepped up to the raft. "My name is Kaz," she told us. "I'm co-captain of the *Bottom Time* and the Rubber Duck is my boat." She would be in charge of getting us out to the whales; she would also help us with our snorkeling skills. I sighed with relief at this.

"We only ever approach the females and their calves," she said. "*Never* the male humpbacks. This is their breeding season, and they're busy showing off for the females. This makes them aggressive and unpredictable. The female whales are very approachable, even when they have new calves, but we treat them very gently and respectfully.

"I don't want to hear any shrieks, screams, or loud noises," she said, looking stern. "I know you're excited about seeing the whales, but you need to stay calm and be very quiet. If I tell you it's okay to get into the water with them, you'll just slip into the water. Just slide and glide. No splashing, no talking."

When she'd finished her speech, Kaz's face lit up with a huge smile. She told us that she and Captain Roger were really looking forward to the trip. Sierra, she said, always seemed to find lots of whales. She knew we were going to have a great time.

As I looked around I thought about the treasure ships beneath the waters. In my opinion the humpbacks are the real treasure of the Silver Banks. I was glad to know how protective the Dominican Republic is of them. This was no huge whale-watching scene like the one I experienced in the San Juans, where a hundred boats may be crowded around a small group of orcas. I was quite familiar with orca screams, and Kaz's prohibition wasn't flattering. I remembered the few major screams I'd let out myself when orca-watching out of Friday Harbor. I resolved not to disgrace myself by screaming when we met the humpbacks. I looked around hopefully.

"Where *are* they?" someone asked plaintively, echoing my thought.

Kaz pointed to the horizon. I saw nothing. "Let's go out anyway," Kaz suggested. "You never know."

Only when you are actually in a tiny inflatable raft on the open ocean can you have any real conception of how small it is. The day was delightfully warm and the languid air was like a caress, but the fact remained that we were experiencing this lovely caress on something little bigger than a floating postage stamp.

I had my usual qualms as I stepped onto the little boat. I am not a water person. Kaz is, though, and she inspired confidence as we putted along. I knew she was an expert sailor, snorkeler, scuba diver, and all-around water adventurer. She is also an expert wildlife photographer and was carrying a camera with a lens nearly as big as the Rubber Duck. What could possibly go wrong? I settled back to enjoy the afternoon, even without whales, and was soon lulled into a kind of drowsy bliss.

I was sitting on one of the pontoons, dangling my feet and telling Teresa about my talks with Granny, the orca leader of J Pod in the San Juans. "Then I talked to another orca, a male named Ruffles," I told her. "The orcas are wonderful . . . the orcas have a fantastic family life . . . Granny is the wisest orca . . . " On and on I went, enjoying Teresa's interest in the whales of the San Juan Islands.

Suddenly the sea exploded. A gigantic sea serpent, the stuff of dreams, rose straight out of the water, writhed once and then flopped down, almost swamping the boat. I heard a scream. It was Kaz! She threw the boat into reverse, almost throwing Teresa and me into the water.

We were all screaming, and all admonitions to be quiet were completely forgotten, even by Kaz herself. After a few circles, she brought the boat back to a stop and we tried to calm ourselves. "What *was* that?" I asked.

"A juvenile humpback," Kaz said. "A three-year-old, or thereabouts. But what did he *say*? I know he said something!"

Teresa and I looked at each other burst out laughing. "Enough about orcas!" we shouted in unison.

"What?" Kaz stared at us.

"What he said was, 'Enough about orcas! This is the Humpback Home! I don't want to hear any more about those whales!'" Teresa told her.

"Are there orcas in the area?" I asked Kaz.

"No, not right around here. They hardly ever come in to Silver Banks. But they could very well be waiting just outside the breeding grounds. They watch and wait, and if they have a chance to take a baby humpback, they do."

"It's as if we were sitting around with a herd of sheep, discussing wolves," Teresa said.

"How great wolves are! How wise! How terrific!" I added. "Of course the sheep wouldn't want to hear it."

The analogy was okay, but this was no sheep! A humpback whale, even a three-year-old, is *huge*. The young whale had towered over the boat, and his leap was so vast he almost seemed to fly. He was so close I could see his long fins very clearly. I felt like he'd waved them at me.

"It was totally eerie the way he crept up on us, wasn't it?" I said. "Here we were just sitting around, not even looking for whales any more, and he swam in to do a little people-watching!

He was following our conversation pretty closely, too. Well, I guess if we can watch them, they can certainly watch us."

"Raphaela, did you see that *New Yorker* cartoon?" Bay called out. "It shows a whale-watching boat, and two whales in the water. One whale tells the other, 'They don't have lives, so they watch us.'" He hooted with laughter and the whole boat joined in.

As the Rubber Duck made its way back to the *Bottom Time,* I lay back and tried to absorb my first humpback experience. The emotions had come thick and fast. The quiet bliss of the sea, then the sudden shock of the whale erupting in front of us, and finally the complete surprise of his declaration. I realized yet again that in any interaction with animals, I am never a detached observer, but a participant in a communication process that goes two ways. The young humpback had shown me very clearly that he was a being as sensitive as he was overwhelming.

I had a collection of poems by Mary Oliver with me, and I found one called *Humpbacks.* I love this part:

We wait, not knowing
just where it will happen: suddenly
they smash through the surface, someone begins
shouting for joy and you realize
it is yourself . . .

Oliver's right. Of all the emotions you feel in a whale encounter, the strongest one is always joy.

Monday, March 22

Today was my day to confront snorkeling. My hastily assembled shopping bag of equipment was not a complete disaster, but it did present some difficulties. Kaz laughed at me when I put on the face mask upside down, with the price tag still dangling from the strap. I don't like putting my face in the water. I'm afraid of losing my contact lenses, and my flippers are too big and give me cramps, which is how you know they are too big.

Water is forgiving, though, and with some patient coaching from Kaz I was soon lying in the water chatting with some fish. The little round orange people were especially nice to talk to. But even they thought it was funny when I tried to get back into the boat.

I was resting on the deck and mentally comparing myself to a beached whale when Teresa came up to me. She wanted to know if I was ready to have a consultation with one of the humpbacks. I was delighted. For me, having a telepathic session set up by another communicator is a real luxury. It's like a masseur getting a massage from another masseur—you're so aware of what's being done for you, and you love every second of it. With Teresa doing the communicating, I wouldn't have to delve for the telepathic link. I could just sit back, ask whatever I wanted to and bask in the whale's wisdom.

I'd been thinking about what I would like to ask, but now, having seen the magnificent whale yesterday, I knew I had to ask the deepest questions I could find in myself. I wanted to know what it was that brought me to this magic place. What could I bring to an encounter with a humpback? I finally decided that the best thing I could ask one of these great beings was how I might best use my skills as an animal communicator for the good for all.

We retired to a quiet spot, and Teresa sent out her request for a spiritual conversation with a humpback. We were soon in contact with a humpback spirit guide. He told us he was not currently in a whale body, but was resting between lifetimes. His calm, beneficent presencc filled my mind with peace. I felt rather as I felt with Ruffles, the incredible male orca with whom I had been half in love during our time with Granny—totally comprehended and accepted in my deepest being.

I asked the guide for help finding my place in the community of telepathic animal communicators. Here on the *Bottom Time,* with nine communicators, all friendly but still competitive beneath the surface, the question had arisen naturally. The two official communicators, Teresa and Penelope, are admired mentors and had been role models. Yet their ways are not my ways. I am an established communicator myself. My consulting practice is very active, and I have many opportunities to help others comprehend animal sentience through my writing and speaking. Does this mean I am competing with other communicators? How could I maintain close ties with them, while also being true to myself and my vision?

"I understand what you are asking, "the whale guide said quietly, "and I can help. First, know that the circle of communicators is huge. There is room in it for all. I know the feelings of competition and envy, for I have felt them myself. When I was in a whale body, I was not one of the largest males. I didn't get many breeding opportunities. In the end I simply had to let all that go. I had to learn self-acceptance. Once I learned the right value for myself, a very high self-value, I became completely restful in myself. This is what you must do.

"Envy is the feeling that comes when you believe you want something someone else has. You don't really want it! Be careful. You are very good at bringing in what you want."

When I heard this from the guide I recognized its truth. I had magnetized a successful consulting practice with very little outer effort; I had simply drawn it to myself. The whale guide's mind reflected back to me a clearer picture of my abilities than I had previously had. My questions about competitiveness dropped away from me. I knew that my real question was not how I would compare to others, but how I should handle my own gifts.

The spirit whale went on, "Be careful of power. Be careful of notoriety. The quality of your work is good. Your clients receive great benefit from communicating with animals through you. They come to you for animal wisdom, and you make it possible for them to receive it from their own companion animals, from the wild animals around them, and indeed from animals everywhere. This draws more people to you, because it is a genuine service. In time you will become famous for your gentleness and your loving kindness in giving this gift.

"Your work will make more people think about the way animals are treated. Many are now doing this kind of work. Before the end of your lifetime, I foresee that huge numbers of people will understand the interrelatedness of all life."

It suddenly occurred to me to wonder if he had ever been a person, and I passed this question on to Teresa.

"Never!" came his forceful reply. "Always a whale. Being a human is . . . " (he seemed to grope for the right word) "a *bitch*! Yes, a bitch, that is right, is it not? Very difficult. A lot of trouble. Whereas a being a whale is a life of ease. I can recommend it!

"You humans! You struggle constantly. I can tell you that in your lifetime many human structures will fail and be reorganized. Even now this is happening. Before, you did not show love for us. Rather the contrary. Now, you come out to see us with such love, and we love it when you come. When you approach us in your little boats, the water actually changes. It changes its life force, or I could say, it changes its sparkles, the little points of light everywhere. You come out, and you squeal, and we squeal too.

"Have you ever noticed that very few of us come to your boats? Out of thousands of us, you only ever see a few. Do you wonder why? I will tell you.

"Some of us have trouble coming near your boats, just like you have trouble with your appliances." I received a picture of me through a whale's eye, fumbling awkwardly in my snorkeling gear. How embarrassing! "That's one reason, but not the main one. Most of the whales don't come because they don't believe it is possible to communicate with humans. They doubt you are capable of feelings; or, if you have feelings, you don't have the finer feelings that we have."

"You only see a very small number of us, too," I responded. "Many people doubt that whales can communicate, just as you doubt it about us. That is changing! I hope it changes among whales, too."

"It will take time, but it will change," he said. "As a matter of fact it is part of my mission to educate other whales about humans. So we are not so different after all."

As the session ended, I found to my surprise that I was putting the very same questions to Teresa that my beginning clients always ask me. "How did you do that? Was it real? How can I be sure that was really a whale guide? Are you positive we weren't making it up?"

When I realized what I was doing, I had to smile. These doubts are natural, and the nature of telepathic communication may even make them inevitable. It is so very close to your inner self, to your mental processes, that you can't help wondering if the communication was anything other than your own thoughts. To be sure of telepathic communication requires constant self-examination. I knew it was real, but I also knew that the need for inward growth and clarity is never completed, and is required of us all.

Tuesday, March 23

Snorkeling is getting a bit easier. Today I finally felt I was competent at clearing my mask, which has been one of my biggest worries. I still feel like a beached whale when I try to get back in the boat, but at least I've started to feel more secure that my contact lenses won't go floating away. That's a relief, because today I was actually in the water with the whales!

I got into the Rubber Duck with Sierra. Kaz has instructed Sierra to keep an eye on me whenever I'm in the water, so I need to stick with her. I guess I do need a baby sitter, and Kaz is a great one. Anyway, I was so glad I was in her boat. I can hardly believe what happened. No one can. It was really incredible.

I knew that Sierra had a tremendous affinity with dolphins. They are a passion with her, and she has even named her travel company The Divine Dolphin. I didn't think we'd see dolphins on this trip, though, because we had been told that they are *never* seen in these waters. Make that *hardly ever!* We were putting along in our little boat, when suddenly a group of the speediest dolphins you ever saw came whizzing right up to the boat. They just appeared in a flash, like a magician's trick: now you see them, now you don't. They buzzed the boat, looked right at Sierra and seemed to be greeting her, and then sped away. We all saw them, although personally I think they may actually have been dolphin spirit guides. Either way, what an experience! Sierra was so happy, it was a joy to see her.

Kaz stopped the raft soon afterwards near a group of three whales: a mother, her calf, and a third whale. She told us that mothers and their infants almost always hang out with another whale who acts as their guardian. It may be a breeding male or a juvenile; sometimes it's a cousin of the infant, or one of its older siblings. When you approach the mother or the calf, you will often find yourself eye to eye with the guardian, who demands, "And your business is?"

The three of them were very quiet, just floating in the water together, barely moving. Kaz gave us the okay, and one by one we slipped over the side of the boat to join them. There were no screams this time. This was the experience I had been waiting for, and it was as fulfilling as I had hoped.

Yes, I was overwhelmed by their vastness, and yes, I was afraid. Yet I was far more overwhelmed by the sense of peace that emanated from the souls of the three whales. Yesterday my whale guide had said that whales live a life of ease. Today I experienced that sense of ease for myself. It's beyond any ease I could have imagined. It's an expansiveness, a sense of oneness with the great ocean, with the universe itself. We are all carried in the ocean of the womb, and perhaps we humans feel stranded and exiled in the air. Whales are born from the ocean to the ocean. Now I was back, too, floating in the sense of oneness that came from them.

Off on the horizon we spotted a group of male humpbacks sporting. "Look at those rowdy males," Kaz said. We got back in the boat and went toward them. A hundred feet or so away from them Kaz cut the engine, reminding us that you never want to get too close to the males during breeding time.

It turned out to be a group of three males. They were lying on their backs waving their long, fringed, winglike fins. They do this as a way of dispersing heat from their bodies, and this is what allows humpbacks to stay cool in these subtropical waters, where every other species of whale would perish from the heat. Even though I knew this, I couldn't help seeing their languid waving gestures as beckoning us to them.

You don't want to go, though, because they are capable of bursting out in lively breeding displays at any moment. So we stayed well away, and just watched. They were on their backs, and we could see their white undersides, against which their penises (which everyone calls "sea snakes") showed bright pink.

Their wild antics, incredible bursts of acrobatics in which they leap out of the water over and over again, and also their incredible songs, are the displays they put on for the females, to entice them to mate with them. They are completely in the grip of their hormones.

As we watched, the three whales began a fantastic display. It was totally gripping. I wanted to see if I could talk to them, but conversation turned out to be impossible. All I could hear from them was, "Choose me! Me! Me! Look how I jump! Look how strong I am, how agile, how brave. Hear my wonderful song!"

I knew it wasn't me they were talking to, so I checked in telepathically with the female we had just left. Was she impressed? If so, she didn't quite want to say so. Her attitude was more like, "Yes, you're great. Call me when you're finished."

"They do whatever they can, but I already know whom I want to be with, " she said. I was left stirred by a sense of mystery. Was she most drawn to the whale who leaped the highest, or the one with the sweetest song? I seemed to hear Fitzgerald in my mind, telling Gatsby:

Then wear the gold hat, if that will move her;
If you can bounce high, bounce for her too,
Till she cry "Lover, gold-hatted, high-bouncing lover,
I must have you!"

Tuesday evening, midnight

I'm not the only one who's feeling whale magic tonight! I was up on deck, quietly watching the waves from a dark corner.

It didn't take me long to figure out that if I waited here long enough, I was going to see everybody on the ship, in every possible combination!

Is it hormones in the air? I hadn't imagined there would be so much . . . animal communication going on. I'm fascinated.

Wednesday, March 24

This trip has now been officially designated as one of the most incredible trips Captain Roger has ever had. Yes! He told us today that he has never seen anything like the display we enjoyed today. Well, actually, he said he did see something almost as mind-blowing once before. To tell the truth, I wish he hadn't told us. I feel like I could have lived without knowing that once, when some passengers were in the water with a whale, it came to the surface right under them without any warning. People were falling off the whale right and left. I don't want to think about that too much.

Anyway, our experience was a hundred times cooler. It took place in the Knife. I like the Knife: it's goes faster than the Duck and it has a canopy, so you aren't constantly exposed to the sun.

I was with a pretty rowdy group, and I couldn't help recalling some of the unusual combinations I'd witnessed last night; I was seeing some of my fellow passengers in rather a new light!

Spirits on the Knife were running high—except, I should say, for Penelope Smith's. She is a staff communicator on the trip, but I think the whole voyage had been difficult for her. Penelope has a love for whales that is truly profound. She has told us that in the presence of whales she experiences the divine, totally and absolutely. Her attitude is extremely reverential, and I suspect she finds the screams of joy and delight from the rest of us, and the really funny remarks the other communicators have heard from the whales, disrupt her worshipful feelings. "Enough about orcas!" isn't quite the profound kind of remark she came out to hear.

Penelope and her friend Odette were sitting in the front of the Knife, carefully maintaining their Sunday School deportment. They'd brought some little wind instruments with them, ocarinas, I think they're called, and they were playing gently on them, perhaps in hopes of creating a more spiritual feeling in the boat. I could have told them it was not going to happen today.

After motoring along for a short while we came up to a mother whale and her tiny baby (tiny for a humpback, that is). Roger said she looked to be about two or three months old. She was precious. We were just admiring her when, without warning, she absolutely *erupted* from the water. We all gasped at her first leap. And her second. And her third . . . and then we realized she was not going to stop! She was going to entertain us for as long as we were willing to sit there and be entertained!

This baby was the poster whale for humpback high spirits. The mood on the boat grew higher and giddier as we were treated to a display of every acrobatic maneuver in the humpback repertoire. It was absolutely fantastic. She breeched, spyhopped, leapt, and twisted in the air. She dove under the Knife and appeared on the other side, only to leap once more into the air. We were screaming with sheer pleasure. Finally Penelope and Odette succumbed, and started laughing, then screaming along with the rest of us. The baby was irresistible. I have never in my life seen such joy in movement, in strength, in life. And from a two-month-old, yet!

I tuned in to her mother, and found her laid back and amused by her energetic offspring. "Let her enjoy herself," she told me. I was more than willing.

The display lasted for three hours! It only ended because we, exhausted, decided to go back to the *Bottom Time*. Checking in with the little whale, I heard her say, "Oh no! They're leaving! What can I do to keep them interested?" Her invention never flagged, and she didn't even look tired! She adored the audience. This whale was born for show business.

Thursday, March 25

Today was the very best day of all, and one I will never forget.

I chose to go out in the Rubber Ducky; it is such a dear little boat and I wanted to be with Kaz and Sierra, my guardian and baby sitter. We putted out looking for whales, and soon spotted two male humpbacks, one obviously a juvenile accompanied by a bigger male, probably his older brother.

I never thought we would be getting in the water with them: Kaz has always been extremely firm on that point. Yet these males exuded quietness. They surfaced a few times close to the raft, then dove, maintaining their position only 20 or 30 feet from us. We stayed near them and as the minutes went by and they continued quietly floating near the boat, Kaz finally said that Sierra could get into the water if she wished. Quietly, Sierra slipped in with the whales. We all felt the atmosphere become even quieter, as if a deeper level of silence had descended on the waters.

Kaz nodded at Teresa, who slipped into the water near Sierra. The two of them floated above the whales. One by one Kaz allowed the other three people in the boat with us to go in. I was the last, but finally she nodded at me. I joined the others in the water.

What it felt like was direct contact with another mind, such as I have never experienced before. The whales were with us as much as we were with them. I looked a whale directly in the eye, and he gazed back at me with total attention. I felt I was being beheld by a mind so vast it was truly beyond my comprehension.

The whale wondered at me in turn as his gaze lingered. I felt him asking, "Who are you?" He was groping to understand me, just as I was reaching out to understand him.

We stayed as still as we could, floating on the surface, while the whales circled and floated below us. Another whale came, looked at us and swam away again; the two whales with us just stayed, barely 15 feet beneath us in the buoyant clear water.

Then one of them spoke. "You could come down to us," he said.

Teresa, Sierra, and I all heard it very clearly. I have never had such an invitation, never felt such a mixture of fear and longing. None of us moved. Afterwards, exchanging impressions in the boat, Teresa said that she had replied, "God, you have no idea how much I would love to, but I can't." She felt she was not capable of sustaining such a dive. Sierra said, "I could." She is an expert diver who can easily go down 25 feet; ten or twelve feet would have been easy for her. She didn't do it because she wasn't sure she'd heard the message correctly. "What if I heard him wrong?" she reasoned with herself. "If I scare him away, the others will be furious!"

As for me, I knew there was no way, absolutely no way. I was terrified. Come down to him? Sure! Can't you just see me? I must have sent back a picture I'd formed based on Captain Roger's story of people falling off the whale because the whale said, in a gentle voice, "We see you, small people. We won't hit you!" I felt the love, the interest, and above all the invitation. If only I could have accepted!

This experience was the one that I will carry away with me from this eventful journey, and I have to say that it is an experience available to everyone. Being a telepathic communicator did certainly enrich it, but I'm quite sure that anyone who had been in the water with those whales would have had a similar experience, regardless of their telepathic training. I have now felt the mind of a whale, and it is very strong indeed. If a whale looks you in the eye, your mind will open to it. This I am sure of.

Swimming with humpbacks is and should be a rare and special experience. We do not permit it in the United States, through fear of disturbing or harassing the whales. The Dominican Republic permits it but only on a very limited basis, thank heavens, and I do not believe the whales are in any way disturbed. After all, if they want to leave any encounter there is no way we could stop them. As the whale guide said, the majority of whales do in fact just leave the boats alone. Amazingly, a few very special, loving, open-hearted whales believe that we humans are worth at least investigating. For that, I will be forever grateful.

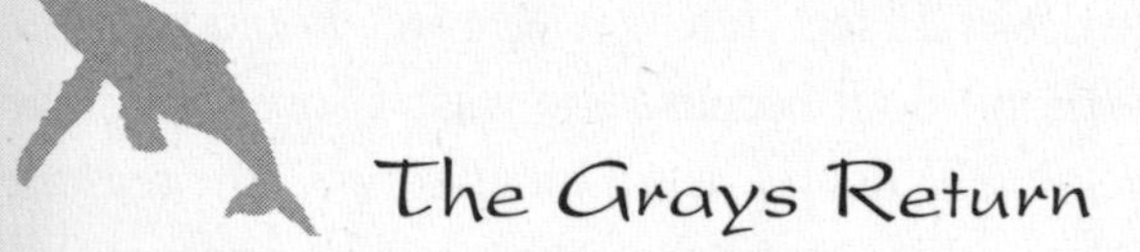

The Grays Return

I spotted an article by Mary Getten in the January—March 1999 issue of *Species Link* a month or two after Raphaela's return from the Caribbean. In it, Mary thanked all the communicators who had taken part in warning the gray whales, and expressed her gratitude that they had escaped harm.

Then she went on to remind everyone that the grays would be returning to Alaska during the spring, and asked the communicators to renew their warning. However, by the time I read the article, it was already too late. Sadly, a three-year-old gray whale had been taken by the Makah.

Once again I called Raphaela. She told me she had already discussed the hunt with Mary, who was very much saddened but who was trying to put the death into perspective. Mary reminded Raphaela to have faith in the whales. Simply put, whales know what they are doing. You can't second-guess a whale, and there are no accidents. Perhaps the whale had a reason for giving herself to the Makah.

I told Raphaela about an article I had found on the Internet about the hunt. It had been posted by one of the *Sea Shepherd* conservationists, and it made painful reading. In theory, the Makah hunt was supposed to be "traditional," and the whale was supposed to be held sacred. According to these witnesses, there was nothing either traditional or sacred about this hunt. The whale was killed not from a canoe but from a powerboat, not with a native harpoon but with a gun. The so-called Makah traditions were simply not there. How could the whale have offered herself to the Makah under these conditions?

"We can speculate all we want to," Raphaela told me. "The only way to know for sure is to ask the whale who was killed. It's always a safe guess that things are never as simple as they seem at first. For instance, did you know that when Mary first contacted the grays, they didn't believe her warning?"

"Tell me more," I begged.

"Well, Mary's first concern when she heard about the Makah's permit was for the resident gray whales who live near Neah Bay. The Makah only had permission to hunt migrating grays, but Mary figured that any gray moving in the right direction would be vulnerable. So she contacted the leader of the resident grays. That whale essentially told her she was crazy. She didn't believe that anyone would try to harm her, and she told Mary so.

"That's when Mary got the idea of having a lot of communicators send out the same message. As she said, if someone calls out of the blue and says your house is about to burn down, you'd think that person was nuts. If 50 people called and said the same thing, you might start to take it seriously.

"Mary sent out a letter to 60 or 65 communicators—every one she was aware of. She asked them to warn the whales, the whale angels, the whale Deva—whomever they thought they could contact, in whatever way they thought would work. Mary then worked on contacting the Deva of the migrating gray whales. A Deva is a sort of over-spirit who takes care of a particular group of people or animals. The gray whale Deva seemed to take her seriously, and promised to keep the whales safe by having them stay off-shore."

"Are you saying that the Deva believed in the danger, but the whales themselves may not have?"

"Possibly, but I don't really know. I took part in the warning effort myself, but I never talked directly to any whale. I just did a sort of broadcast. You know, 'Calling All Grays.' Whether they believed the message or even received it, I just don't know. Maybe it was the Deva who kept them off the shore during their southern trek. Who can say for sure? Maybe it was just a lucky coincidence. Maybe when they escaped any harm on their southern migration, they concluded that there was no real danger after all. Let's not jump to any conclusions. Let's ask. I'll try to reach the whale so we can find out for sure."

Raphaela centered herself and sent out a telepathic call. She called the whale Yabis, which is the name she was given by one of the Makah elders. To reach a whale telepathically it helps to have some way of identifying her, so Raphaela asked for the whale who had been killed in Neah Bay. Even though the whale was not in its body, she thought she should be able to reach its spirit.

I sat quietly waiting on the other end of the phone line while Raphaela sent out the telepathic call. She had allowed me to be the "interviewer"—in other words, I would suggest the questions to ask the whale, while she kept the telepathic link open. Raphaela and Mary had evolved this method when interviewing the orcas because they found it difficult to maintain the quiet, open state of mind required for telepathic communication, and still focus on the outer world clearly enough to conduct an interview. This was my first time to participate in a whale interview.

After some time, Raphaela came back on the line and told me that Yabis had responded. It seemed she was a modest spirit, very quiet and retiring.

"Tell Yabis that we are writing about the grays, and we'd like to learn more about the hunt from her point of view," I said.

"Yes, okay," Yabis said faintly.

"Ask her if she experienced a call from the Makah," I said. In traditional hunts, we had heard that a call goes out for an animal to offer itself. I wondered whether the sacred bond we had heard of between hunter and hunted actually existed in any form in this case.

"I did hear a call," Yabis said. "It wasn't directed particularly at me. It was just a call, and then I was the one who answered."

"Ask her if she knew why they were being called," I said. "When she approached the Makah boat, what was she expecting to happen?" I was afraid Yabis may have thought it was like one of the whale-watching boats that regularly approach the grays in the bays of Baja California.

Yabis didn't answer at first. She was reluctant to go back over the events of that day. "It doesn't matter," she told Raphaela. "In the grand, global scheme of things, the death of one little whale isn't very important."

Yabis then showed Raphaela a picture of herself as one tiny speck of light in a firmament filled with other tiny specks. She was in her cosmic afterlife, surrounded by other souls; at the same time, she seemed rather solitary. She was loathe to go back and involve herself again in the world. "Don't take me over all that again," she implored.

Raphaela gently pressed her. "Please, Yabis," she said. "We really need to know.

"No," Yabis finally said. "I didn't know why I was being called. I was curious about the people. Who were they? What were they doing? That's all."

Raphaela said, "I'm getting that when she was killed, her reaction wasn't total outrage. She didn't seem to be saying 'No, no, this is totally wrong.' She was surprised and shocked, of course, but she had a kind of sad but resigned acceptance."

"Ask her if she is aware of any tradition of whales being hunted in her family," I suggested.

"She's showing me that she knows what hunting is all about." said Raphaela. "When she hears the word 'hunter' she doesn't think of humans first, though. She thinks of orcas! Orcas do hunt baby grays, and that is their biggest danger. Then there are ships—being hit by ships. Then she thinks of human hunters, but that's last, because it hasn't happened in so long. All these things are the same sort of danger, she feels. All part of 'being hunted.'

"As for having some kind of mystical bond with the hunters—she says absolutely not. They don't have a relationship with any hunters. They certainly don't regard themselves as existing to provide food for anyone. 'We were not put here in the service of hunters,' she says. 'That's not how we conceive of ourselves!'"

"Ask her about her family. How do they feel about it?"

"She's showing me that her family is mainly her mother. She had some older siblings and a younger one, but she wasn't always with them. She's showing me her family as a sort of a net, with a gaping hole in it. The hole is her—her absence from the family. Energy drains out of the family through this gap. But it's being patched with something translucent which is helping hold in the energy. I don't quite follow what she's saying. It's whale philosophy."

I thought I understood that Yabis was showing us the healing her family was experiencing after their loss of her.

"Ask her about last fall," I said. "Did the grays hear all the communicators calling them?"

"Hey, that question really got her attention!" said Raphacla. "She says yes, they certainly did! Yabis has suddenly become much more interested in this conversation. She probably didn't realize we were connected with the contacts we had last fall. She's telling me that they all just loved hearing the voices. They were intrigued, and it made them feel terribly important. All those people calling them!"

"Did they believe the warning?"

"She says, 'Some did and some didn't. We were all talking about it. We were fat then. We had been eating well all summer, we had lots of energy and we could afford to take detours if we wanted to. So some people decided to try it.'

"Of course!" Raphaela said. "That's another explanation for the behavior that we hadn't even thought of. Gray whales get most of their food during the summer. They feast on all the rich plankton blooms in the cold waters off Alaska. During the winter, when they're down south in Baja breeding and having their calves, they hardly eat. So they're much thinner on their northern migration."

"That's right," confirmed Yabis. "Going back, we were thin. We needed to take the direct route, close to shore. We were hungry. Maybe if we'd heard voices again, like we did in the fall, some of us would have tried to stay farther out, but we didn't hear anything. And anyway, we might not have been able to. We had to move quickly. We needed to get back to our food."

We'd heard from Mary that Yabis, and also other grays who had died of natural causes during the spring, had been unusually thin. "Ask her about that," I said.

"She says they were maybe a little thinner than usual, but she doesn't know why."

I thought about the calculation a whale might make under such circumstances: should she stay close to shore, and perhaps risk a meeting a Makah hunter, or should she take the chance of swimming father out to sea, which would take more energy, thus running the risk of not making it to the rich Arctic feeding grounds? The choice was not so simple, after all.

"Ask her about next fall," I said. "Will the grays remember about the danger on their next southern migration and stay off shore?"

"She's telling me that whales migrate at different times. There are still stragglers making their way north when it's time for the early whales to leave again for the south!"

"What does she recommend that we do, then?" I asked.

Raphaela laughed. "Yabis just reminded me that she's not there!" she told me. "Then she said, 'Remember, losing a whale or two is just not that big a deal.' She has an amazingly cosmic perspective about the whole thing."

I appreciated that, but she needed to know that we weren't just talking about one or two or even five grays taken by a Native American tribe. "Tell her that we're worried that it may open the door to a return to more active whaling. Hunting of whales on a large scale, like in the old days."

"She understands." Raphaela reported. "She gets it now—she didn't realize all the implications at first. She told me, 'Yes, people do need to be reined in.' What a sweet way to put it. Whales are always so much kinder and gentler to us than we deserve.

"Anyway, she says we will need to tell the grays again this fall. She says, 'You'll need to remind them again and again. They are coming and going almost all the time. Just remind and remind.'"

"Ask her if there is a special whale we might talk to. I don't know if that would be better than a general broadcast warning."

"She says what you did before was very effective. Do that again! But if you want to talk to an individual whale, you can call her mother. Wait, she's telling me her mother's name. No, I can't get it. It's one of those musical whale names . . . they've very hard to hear."

"Try again!" I was thrilled by the idea of hearing a whale's name directly from another whale. "Tell her to try sending it to me, too. Maybe I can get it."

Raphaela put down the phone and got very quiet. I did the same. I had been clinging to Raphaela's words and had felt Yabis' presence coming in more and more strongly as the interview went on. Now, I was almost hearing Yabis myself. I am always more telepathic when I am around Raphaela. I opened myself to her as fully as I could, and listened carefully for the name.

After a moment, I heard a little humming sound and then a gentle explosive sound, like a whale's breath, and then some more humming sounds. I scribbled "Pianiyah" on a piece of paper.

"What did you get?" I asked Raphaela.

"This isn't quite right, but it started with an M or a B. Something like Bianca."

"Bianca...Pianiyah. Very close! I heard a humming sound too—that was probably the 'M' you heard at the beginning. B and P are very similar sounds. I think her name must be something like Mbianiya."

When I said this we both felt an almost overwhelming flow of blessings streaming to us from Yabis. It was as if her mother's name had been a mantra which drew us to her. As we approached the heart of the whale, the tiny point of light Raphaela had seen at the start turned into a glowing star. This beautiful soul, who had accepted our first questions so quietly and modestly, now allowed us to glimpse her in her true nature. It was completely beautiful.

"I will help you," she said, speaking directly to both of us. "I will be a lighthouse for my family. You will be lighthouses too."

That is what we must all be, for the grays and for all whales.

We ended the interview full of gratitude to Yabis, and immediately began making plans for regular contact with the grays. Any reader who would like to help is very much invited to do so. You need only add the name Mbianiya, gray whale, mother of Yabis, residing off Kodiak, Alaska, Baja, California, and all watery points in between, to your prayers. If you feel you can send a telepathic message, please do, and remind her and all her family to stay away from Neah Bay, on the northern coast of Washington.

The ban on whaling must not be weakened. It must be made absolutely as strong as possible, so that whales can continue to bless our lives on earth forever.

Chapter Five

Playing with Dolphins

Bottom Time Again

Several months had gone by since Raphaela's return from her humpback adventure. She had settled back into her daily routine: consultations with dogs, cats, rabbits, birds, and horses—the endlessly fascinating animals whose lives, intertwined with ours, mirror back to us our true nature as part of the beautiful fabric of all beings.

The humpbacks were still with her, forming an enormous background screen on which the images of the animal companions she spoke with each day played. Size, of course, is not the measure of an animal. The soul of a cat or a dog is as immeasurable as the soul of a whale. Yet when you swim with a whale, you can hardly escape receiving an overwhelming impression that they did not come by their hugeness by accident. It is part of their being, and implies a certain majestic serenity in their souls as well.

Raphaela was still immersed in these impressions when Sierra called her again, with another opportunity to spend time in the world of wild sea creatures. This time, Sierra was organizing a dolphin-viewing trip to the Bahamian Islands. The Caribbean around Puerto Rico and the Bahamas is prime dolphin territory. Atlantic Spotted and Bottlenose dolphins live there in huge pods, thousands of dolphins strong. Would Raphaela like to join this trip, this time as the "official" animal communicator?

Raphaela didn't hesitate for a minute. She knew that dolphins were Sierra's first love, and that she felt a special affinity for them. The dolphins obviously returned her affection—Raphaela remembered the flying visit a group of dolphins had paid to Sierra on the Rubber Duck. A dolphin trip with Sierra would be very special.

Dolphins are, in a sense, founding members of the animal-communication movement. Humanity's association with dolphins goes back at least to the beginning of modern Western history. There are legends from ancient Greece and Rome of people being saved by dolphins, riding on the backs of dolphins, fishing with dolphins. Dolphins are sacred to Aphrodite and draw her seashell cart through the waves.

In our own day, dolphins and people have drawn even closer. There's just something about them—their playfulness, their charm, their obvious intelligence—that attracts us, and the attraction seems mutual. Almost everyone who has associated with a dolphin has had the feeling that the dolphin is positively eager to communicate. Whether you're into telepathic communication or not, you can truly almost hear a dolphin thinking.

Animal communication is a two-way street. Telepathic communication works best when the animal on the other end wants it as much as you do. Clearly, dolphins have been trying as hard as humans to establish a telepathic link. Like a tunnel being dug from both ends that finally meets in the middle, the result was today's exciting explosion of interspecies communication.

Did she want to go visit the dolphins? Of course she did.

Raphaela called and asked if I wanted to go too. I would have adored it, but once again, commitments kept me in Eureka. "Be sure to write in your Journal," I begged. She promised to write faithfully.

Raphaela's Journal: Sunday, June 27, 1999

What joy to be back on the *Bottom Time* again! There's almost the same crew as we had on our trip to the Silver Banks. Captain Roger greeted me like an old friend, and Kaz looks as lovely, strong, and competent as ever. I've already settled back into my little stall. There's one big difference from last time, though. Since I'll be doing telepathic consultations for all of the twenty-plus guests, this little bunk room will have to double as my consultation room. Let's see, there just might be room for one chair. The guest can sit there, and I'll crouch on the bunk. Tight, but it should work.

Kaz gave us an introductory talk after dinner. The good news is, this trip doesn't begin with a six-hour high speed trip like the one we suffered through last time. We just start out tomorrow morning putt-putting along. We all stay on the *Bottom Time,* rather than going out in groups in the Rubber Duck or the Knife. All we do is putt around and wait for the dolphins to come and play. Naturally, we try to make it as easy as possible for them to find us.

Kaz was very specific about the rules. The dolphins have be in control of all our interactions. They come and go as they wish. If they come up to the *Bottom Time* we will be able to get into the water with them. But we absolutely can't ever swim after them or try to touch them. If they want to stay and play with us, well, they just will, that's all.

Swimming with dolphins! I'm totally excited. I spent hours in the library before the trip, boning up on the two species that live in these waters. Bottlenose dolphins (like Flipper) are much larger, and weigh about 300 pounds. The Atlantic Spotted dolphins are smaller, speedier, and very zippy. Both kinds are incredibly smart (smarter than we are, lots of people think). New Age types say they are messengers from another planet, enlightened beings come to spread wisdom and peace, cosmic rays of light, and lots more. I thought that made them sound a little intimidating, but what everyone says is, dolphins are pure play. Pure fun. Now, that's a message I can live with.

Monday, June 28

The dolphins showed up right on schedule this morning. A group of ten or twelve Atlantic Spotted dolphins turned up about ten this morning. Everyone says they are just as zippy and playful and fun as we had heard they would be. I was down in my little hidey-hole doing consultations, so I didn't get to see them, but I heard all about it at lunch.

It seems the *Bottom Time* was just kind of bopping along when the dolphins suddenly appeared near the bow. Kaz said, "You know, these dolphins really love our boat. They seem to look out for us—and they always show up about this time. After a late breakfast, I guess." Everyone who wanted to was allowed to get into the water with them.

Captain Rog reminded everyone to behave like a "pod." In other words, they were all supposed to stick close together in one group. It's the only way Roger and Kaz can keep track of the bodies in the water. Everyone is supposed to stay on the starboard side, and never to swim to either the front or the back of the boat. Roger and Kaz are extremely careful of our safety and are checking the group constantly: "Sixteen people in the water—do we still have sixteen?"

The sensation of the morning was a guest named Emily. She is Vietnamese, and very tiny, no more than five feet tall, if that. She was the first person over the side, and in a moment everyone saw her paddling quietly and determinedly *away* from the group and *towards* the front of the boat. Roger of course started screaming, "Emily, get back!" But here's the funny part: the dolphins all went streaming after Emily!

By this time the whole group was hollering at her. She finally turned around and looked over her shoulder, absolutely amazed. "Is there a problem?" she asked. She said she was just practicing a breathing technique she'd learned on a yoga retreat. It made her kind of spacey and she just swam without really being aware where she was going. She didn't hear everyone yelling because she was concentrating on her breathing, or maybe because the yelling was drowned out by the noise of her snorkel. Anyway, she was oblivious.

"Stay closer to the group next time," Roger told her, but moments later she was off again, straight towards the front of the boat. And again the dolphins went right after her. She's got a consultation scheduled for tomorrow. I'll be sure to ask the dolphins what's up.

After lunch I got my turn at dolphin play, and it was so much fun. At first I thought I might have missed today's dolphin visit completely. By the time I got up onto the deck after my consultations, around noon, they had all swum away. I had lunch, and then we all sat around on deck talking. About three, to my delight, here came the dolphins again! It was a group of Atlantic Spotted dolphins, probably the same ones who were here this morning. Roger gave the okay and I slipped into the water with the rest of the group.

Talk about contact! This was about as up close and personal as it gets. These little spotted dolphins are physically about our size. A big one weighs about 150 pounds, and they're three to four feet long. They look better in the water than we do, though. They swim all day and they're much stronger and, well, in better shape than we are. That's the nice way to put it. Elizabeth's cat Cocoa would probably just say we're pathetic in comparison.

As usual, I had trouble seeing them clearly. My eyes aren't very good, and I'm afraid to wear my good contacts for fear of their floating away, so I have on an older pair. Then there's the plastic face mask, which is about as clear as my eyes. I hate to admit this, but the first words I spoke to the first dolphin I saw was, "Would you mind coming a little closer? I can't see you!"

The dolphin obligingly swam right up to me. It didn't take clear vision to feel the intelligence that radiated from her. She was absolutely seeking communication, and it seemed she wanted to go eyeball to eyeball. That's not so easy if you have one eye on each side of your head, instead of both in front like us. She kept turning from side to side, looking at me out of each of her eyes in turn.

I couldn't escape the impression that she wanted me to follow her to the front of the boat. She'd give me the eyeball, or eyeballs, then turn provocatively and swim forward. It was an unmistakable "follow me" signal. That's something else I want to ask the dolphins about when I tune in tomorrow morning. What's in the front of the boat? What do they want us to do up there?

After an hour or so we got back on the *Bottom Time,* and Roger said, "Let's show the dolphins a good time." We all went to the bow and lay on our stomachs with our arms hanging over the edge of the boat. Then Roger gradually nudged the *Bottom Time*'s speed up enough to make a bow wave for the dolphins. The dolphins swam right along with us, surfing the wave. It was the most fun I could imagine. They were just having a ball. They surf the wave for a while, then flip over, swim under the boat (I guess—it's hard to follow their antics because they are so quick), then the next minute they're back in the bow wave. We were all laughing our heads off. What a trip!

Tuesday, June 29

Well, I did get to talk with a dolphin this morning. The one I contacted couldn't have been more open and helpful. First of all I got the straight poop on Emily. It's very sweet.

Emily came for her consultation, but before we got to her dolphin questions she wanted me to talk with a beautiful red fox whom she sees sunning himself on her tennis court back home. That was a new one for me, but I'm always willing to try. I called "The fox who sits on a tennis court in Connecticut" and sure enough, he did answer. When I asked him why he always sat there he said it was because he liked it (surprise) and why would Emily ever want to leave?

We moved on to dolphins. I didn't know a specific dolphin to call, so I asked to speak to the "oversoul of the dolphins, or the dolphin Deva, or the people in charge." I thought that would do it. It seemed to work, because a dolphin came in, greeted us and asked what we needed. So I said, "Emily wants to know why the dolphins who visited our boat were following her yesterday."

"Do you mean that child?" the dolphin asked.

"She's not a child. She's a grown woman."

The dolphin was amazed. It seems he and his friends thought she was a child because she was so small and was wearing a life vest. The breathing exercise she was doing made a rather peculiar noise and they were worried that something was wrong. That's why they were following her—they just wanted to help.

I thanked the dolphin for his concern, and assured him that she was fine and he didn't need to worry about her. Now we'll have to see if they stop following Emily. I don't really know yet how seriously they take me. This will be interesting.

"By the way," I asked the dolphin, "I also noticed yesterday that you and your friends were trying to get us to swim with you towards the front of the boat. Is that right? Was I reading you correctly?"

"Yes. We want you to come with us. Why don't you?"

"We're not supposed to follow you. Those are our rules. You have to come to us. We don't want to harass you." The dolphin seemed amused by this. Like we could harass them if they didn't want us to!

"Why do you want us to follow you?" I pursued.

"To have some fun! You saw us playing in the boat's wave yesterday afternoon. We want you to play with us!"

I had to tell them the sad news. We can't do it. We just can't. Their disappointment was palpable, but what could I do? Then the dolphin had some sad news for me in return. We were boring. They liked us, they enjoyed swimming with us, but . . . we were boring. And there it was!

Wednesday, June 30

In my animal communication practice I always find it so wonderfully confirming when an animal and I talk about something, and then it happens. I hold on to those moments of "proof" when I get one of those doubting moments in which I suspect myself of having an overactive imagination. That's when I'll always think of an animal like Bobbie Kitty. When she showed up just where she was supposed to be, she gave such beautiful confirmation that we really had been in communication.

I was thinking of this because dolphins are truly the most responsive animals I have ever talked with. This morning I learned that they had responded perfectly to both of our communication subjects of yesterday. They didn't follow Emily, and they stopped inciting us to the front of the boat once I disillusioned them about our ability to surf the bow wave.

Now that they know us better, I bet they can't believe they ever imagined we could! It stung to be called boring, but the dolphins are right, our swimming leaves a lot to be desired. I wanted to say, "You should see me on a horse!" but it didn't seem to make sense in this context. But we're rallying. We've come up with a plan to become a little more amusing. You know, put on a little more of a show for the dolphins.

I happened to have seen *Dangerous When Wet* a few dozen times, and Esther Williams is one of my fantasy figures. Okay, okay, I know what you're thinking. Maybe I didn't go out for Sea Colts when I was in Junior High, but a couple of the people here do have a little synchronized swimming experience. They've got us organized to swim in a circle, with the people who can do back somersaults and point one leg at the sky do it in the middle. Hey, something just clicked into place. The maneuver that kept me out of Sea Colts was called a Back Dolphin! Now I get it.

"Did they like that?" people yell to me after one of our shows. The dolphins politely tell me, "Yeah, yeah, very nice," but I can tell they respect the effort, not the show. Oh well!

Thursday, July 1

I've settled into a very pleasant routine. Mornings, I'm down in my hidey-hole doing consultations. Then I go up on deck for lunch and hear the morning's dolphin stories. The dolphins are definitely creatures of routine, too. They show up each morning right around ten, and leave about noon. Then at three they're back for the afternoon shift. That's when I get to play with them.

I don't mind much missing the morning dolphin session. Even though it seems like I'm stuck in a tiny bunk room, I'm really not. I'm out in the ocean with them, communing telepathically. Or, if I'm not talking with the dolphins I'm with some other animal, like Emily's little red fox on her tennis court back home.

When I'm in the water with the dolphins in the afternoon, the others are constantly calling out to me, "Ask them this! Ask them that!" They have their favorite dolphins, too. There's one we all call Stubs who has a missing dorsal fin. I asked him about that, and he said he was in a shark attack. Gosh! It didn't seem to have traumatized him particularly, though. He's as playful as the rest of them.

I asked one dolphin how it was they had so much time to goof around with us. He showed me that a strong, active dolphin can get all the food he needs for a day in about twenty minutes of hunting. In other words, that's his work day! Then he's off for the rest of the day. Whales are playful too, but it takes a lot of krill to feed a humpback, so their workday is longer. With dolphins, play is what they *do.*

Today we had another demonstration of responsiveness from our dear dolphins. It also showed us all how truly caring the dolphins are. We spent last night anchored at a little port called Bimini, which we'd put into in order to clear Bahamian customs. Bimini is a Hemingway place. He used to hang out in a bar there and talk fishing; it's decorated with huge pictures of Hemingway and fish. The whole group of us hit the place yesterday evening and made a night of it, dancing, drinking, and living it up, inspired by the example of the carefree dolphins.

We found out there were other dolphin boats in port, and we fell in with some of the people on a boat called *Dolphin Spirit.* Amazingly, they had been out for a week and hadn't seen a single dolphin! Of course they were envious when they heard hear that we've seen them twice a day.

One of my old students from California, April, was quite upset on their behalf. She is so into dolphins that she actually moved to Florida to be able to spend more time with them. She came to me this morning and said, "Raphaela, you have to do something!"

I said, "April, you're an animal communicator too—you do something!"

So this morning at breakfast April told everybody about the *Dolphin Spirit*'s strange lack of dolphins. She asked the group to join her in a prayer, which she led as follows: "Dear Dolphins, we love you. Please don't leave us! But we know there are thousands of you in these waters. Could a few of you go visit the *Dolphin Spirit* and give them the joy of your presence?"

Everyone joined in the prayer. Then I went down to my hidey-hole and heard no more until lunch time. Then Captain Roger told us he'd heard on the ship's radio that the *Dolphin Spirit* had dolphins by eleven this morning.

They must have heard April's prayer, and just said, "Let's go."

Friday, July 2

Sierra told us today that she has bought some land in Costa Rica. It's waterfront property on the Pacific ocean, near a wildlife preserve. There is a lodge on the land, and she plans to develop the property as a place for her and her guests to hang out with wild dolphins. It sounds utterly wonderful to me, but people have told her that she won't be able to swim with the dolphins around there. They are too wild.

She was undaunted. She's never met a dolphin she didn't like, or who wouldn't swim with her, so she decided to consult with the real experts, the local dolphins she knows so well. While we were all in the water with them this afternoon she asked them how she could get to know the dolphins of Costa Rica.

"Don't worry, we'll help you," came the response. Then Sierra felt a suggestion from them that she give them something. She went below, got a scarf of hers, and tossed it to one of the dolphins. He tossed it gracefully to another one, who swam with it for a few moments and then passed it on to yet another dolphin. They did a whole ballet with the scarf. When they were finished they passed it back to Sierra and said, "Give this to the Costa Rican dolphins. They will feel our good wishes and our love for you."

It touched me to see them do it, and I remembered seeing the Dalai Lama do almost the same thing. By custom Tibetans hand him a prayer shawl at the beginning of an audience. He holds it while he talks with them, then gives it back, blessed. The dolphins certainly blessed Sierra's scarf with their lovely dance. Afterwards Sierra passed the dolphin-blessed scarf around, and each of us in turn held it and added our blessing. I love to think of it going from "our" dolphins to the "new" ones, who I'm certain will soon be Sierra's friends.

At dinner last night we got to talking about all the Atlantic Spotted dolphins we've seen, and someone happened to say, "Aren't there supposed to be Bottlenose dolphins in this area too? I wonder why we haven't seen any." Then we all began to wonder. I was asked to inquire where they are during my morning consultations.

This morning I tuned in and asked for the Bottlenose oversoul or Deva or the people in charge, just as I have been doing, only this time I definitely said "Bottlenose." When I got a response I simply asked whether they would bless the *Bottom Time* with their presence this week, since we would dearly love to see them.

I got back a lot of caution, bordering almost on fear. The dolphin reminded me that Bottlenoses have suffered and are still suffering at human hands. They are the ones that end up in tuna nets, as they are just the right size to be caught. They have been captured in great numbers and many of them are being held in aquariums or are displayed doing tricks. They are also used by the military.

As always when an animal confronts me with the actions of my own species I am filled with shame. I apologized to the dolphin, but it felt pitifully inadequate; nor could I promise that the wrongs will not continue. I know they will. Yes, I could see why they would be cautious.

I told the dolphin that I fully understood the reluctance of his people to come to us. Then I reiterated how much we would love to see them. I assured him we had nothing to do with the terrible things he had described. We were only a small group of people and we all loved dolphins very much. We had been enjoying his Atlantic Spotted brothers all week, and it would be such a joy if we could see them as well. I sent him the love and respect I felt and that we all felt.

My sense was that he accepted my statements and was somewhat reassured, but he said only, "I will take your request to the committee." That was my interpretation—the Bottlenose committee, or something like that.

The committee must have given it a cautious "We'll try it and see," because not long after my request, they came! A small group of Bottlenoses swam up to the *Bottom Time.* I of course didn't get to see them—I was still down in my bunk room, and this time I seriously regretted missing the sight.

We were in about 30 feet of water, but the water is extremely clear in these parts and you can see right down to the bottom. The Bottlenoses stayed very deep, actually swimming *under* our familiar group of Atlantic Spotted dolphins. They stayed there just watching for about ten minutes, carefully keeping the other dolphins between themselves and the ship. At last they must have been satisfied we were harmless, because they began to swim closer to the surface and to interact with the people in the water.

Even so, I was told, their interactions could be described as "judicious." They are of course much bigger than the Atlantics, and a little calmer as well. There was none of the zipping and bopping around that we've gotten used to. Still, they came. I am so happy about it. Animals are almost always so much more forgiving of us than perhaps we deserve. It is such a benison. We all feel completely blessed.

Saturday, July 3

This morning I had a consultation with Wanda, a very sweet woman, somewhat older than most of the other guests, who is here with her husband Lou. I'd spoken with Wanda's dog and her three cats before the trip, when she'd asked me to tell them about the arrangements she'd made and to reassure them that she'd be back soon. This time, she was concerned not about her pets but about Lou.

"I don't know if you've noticed," she said, "but Lou is the kind of person who always puts other people first. He's a giver; he's very modest and he's never pushy. I love him for it, but what's happened on this trip is that he has hardly been in the water. He always waits for everyone else to get in first, and by the time it's his turn we've all had to come out again. He hasn't interacted with the dolphins at all. He hasn't complained, but I know he's disappointed and would love to have had closer contact with them. I was wondering if you could ask the dolphins, as a favor, to do something special for Lou?"

I was charmed by her request, and immediately tuned in. This time I asked to speak to "the dolphins we see every day."

When I got a response, I asked them if they would be able to seek out Lou. What came back was that they weren't sure who Lou was. What did he look like?

"He's an older man. He has white hair."

"We will look," came the reply.

That afternoon we were in the water with the dolphins when Wanda suddenly looked at me, her eyes wide with amazement. "Did you hear that?" she said. I hadn't. She said in a wondering tone, "That dolphin just told me, 'Get Lou in the water!'"

This wasn't the first time another guest had received a telepathic message from one of the dolphins. In fact, it had been happening more and more often all week. Dolphins are so communicative that it takes little more to hear them than some validation that what you're hearing is real. All week people had been saying to me, "I thought he said this," and I only had to say, "That's right."

This was the first time Wanda had received a telepathic message. I nodded encouragingly and said, "Well? Go get Lou!" She hurried up the ladder. A few moments went by and then I saw Lou come over the side and slip in.

All around me I heard dolphins saying, "Is that Lou? Which one is Lou?" I pointed at him. Immediately three dolphins buzzed over and began nuzzling him. I saw Lou break out in a huge smile. A juvenile dolphin joined the group, got right in his face and gave him a wonderful burst of dolphin blips and squeaks, like a song just for him. He beamed even more broadly, if such a thing were possible. The dolphins stayed with Lou for the rest of the session, and Lou was the envy of the entire group.

Wanda hadn't told anyone about her request, for fear of being disappointed. Now, at lunch, she couldn't wait to tell everyone what she and I had done. She was filled with gratitude to the dolphins for the honor they had done to her and her husband. Lou was too. He said he felt like a king.

As for me, all I could think was that this is what it must be like in heaven. It's supposed to be a place of blissful leisure, where your only activity is adoration of the Divine. Dolphins, with their famous 20-minute workday, come close to this ideal of life in the blessed realm. Seeing them in action gives me a feeling of what it might be like, if or when I ever get there. A life of play and giving—that is how dolphins live.

I've put it to myself that they play, but when I think about that word, its meanings start to multiply. Certainly they do play in the sense of sport—their sport is swimming in the most elegant, acrobatic, synchronized way imaginable. They sport about in their element like birds soaring in theirs, nor is there anything you could possibly construe as "work" about it. They're not hunting, they're not chasing anything—they are truly playing. Their swimming isn't a competitive sport (or maybe it is—they do seem at times to be going for the most beautiful, athletic leap any dolphin has ever made.)

I also think of the way Elizabeth uses the word "play"—as in playing her cello. It seems like the very apotheosis of the word when she tells me that her string quartet is "playing" Beethoven, or her orchestra is "playing" Shostakovich.

I have the same fear of heaven many people have—that it will be boring. Sin, we perhaps wrongly imagine, has the attractions of variety and stimulation. Take away lust, gluttony, and pride and what will we have in heaven to keep us interested? Dolphins show us that what we will have is play. Music shows us play that is never dull. It's play that is difficult and technical, and at the same time emotional and immediate. It transcends every paradox—it's there for everyone at every level of life, and it offers riches no genius can ever use up.

When I think about the dolphins, I'm sure that this is the way they play. I saw over this week that everything I or anyone asked of them, they happily did. They came to us every day. When we asked them to go visit the *Dolphin Spirit*, they went immediately. When Wanda asked them to do something special for Lou, their only question was, "Which one is Lou?" They were like angels, going here and there bestowing happiness.

The image that will stay in my heart is of the dolphins playing with Sierra's scarf. I thought at the time of the Dalai Lama and his way of blessing the prayer shawl of his devotees. Afterwards, when I replay it in my mind, I see something else. It was a ballet—spontaneous, graceful, coordinated, musical, and extremely beautiful. The dolphins were dancing to a music that plays to them always, to their inner beings: the music of the oceans, the music of life.

Sierra's New Friends

Raphaela's Journal: December 15, 1999

I got a wonderful letter from Sierra today from her new home in Costa Rica. I had written to ask her how she was doing with the dolphins there. I never doubted she was swimming with them, and I wanted to know if she'd been able to give them the scarf, and if so what their response had been.

Sierra wrote that yes, she had brought the scarf back to Costa Rica with her, and had taken it into the water with the dolphins several times. "The dolphins were very interested in the scarf," she wrote. "I could feel the energy of the Atlantic dolphins merge with the energy of the Pacific dolphins. I brought the scarf in the water with me until it was shredded to pieces and I could no longer bring it along."

There—I knew Sierra would be in the water with them in no time! She went on to write, "Our relationship has definitely grown as we all get to know each other better and develop trust. We have identified several of the dolphins that live here in Drake Bay. We call them Los Primos ("The Cousins"). Our encounters with the large pods of dolphins (often in the thousands!) are more incredible all the time. We regularly have close encounters with them both in the water and out. We have recently begun to kayak with the dolphins, and this is an experience beyond words.

"I will never forget my first kayak with the dolphins. Our cook was hanging out down at the beach one day, watching the activity on the Bay. She came up to the main house and told me that there were lots of dolphins and birds out in front of the point. I hurried down with the kayak and paddled out as fast as I could.

"Of course once you are in the water, the dolphins always turn out to be a bit farther out than they looked from land. I was paddling as fast as I could but still hadn't reached them. I was getting pretty tired, but I could see the dolphins about 50 yards from me. I sent them a message that said, 'you had better come over here to mc because I can't keep paddling like this much more.' In that second at least 25 dolphins turned and swam right for me.

"They came around on all sides of the kayak. I started paddling, and we all moved together as one pod. The dolphins were jumping, diving, playing, fondling, laughing all around me, and I was ecstatic beyond words. Never, except in my very best dolphin swims, have I felt so close and so at one with the dolphins. These were our very own homeboys, Los Primos. You know, the ones who are supposed to be too wild ever to get in the water with!

"I think kayaking is the best way to interact with dolphins. There is no engine noise, it is just you and the dolphins, and you can move almost as quickly as they do with a kayak. They love it!!! And so will you!

"The other day we had a three-hour encounter with a large pod (around 60) of Rough-Toothed Dolphins. They are fairly rare and are seldom seen; very little is known about them. However, they let us into their world for a very long time. They swam very close to us and showed us a wide range of behaviors. I felt they were interacting with us on a very deep level.

"It is such an amazing spot here. So amazing that I have started the Delfin Amor Marine Education Center. Our first goal is to have Drake Bay and the surrounding areas named a National Marine Sanctuary. The Discovery Channel has promised to help us. They are coming with *The Quest,* a 200-foot boat that also holds a helicopter, submarine, and seven smaller boats. They are going to come out in March to help us document the variety and numbers of cetaceans in the area, to help us demonstrate how much this area needs and deserves to be protected.

"This is the most beautiful spot on earth. As I write I am lying in my hammock on the balcony of my bedroom, which overlooks the beautiful rain forest and waves crashing on my palm-tree-lined beach. Out in the beautiful, warm Pacific are countless dolphins and whales, never very far away. Hibiscus flowers sprinkle the jungle with red, and macaws and toucans fly through the trees. I believe I am here to help the dolphins and whales in this area by making it an inviolable sanctuary for them. Dolphins and whales have taught me so much about living with joy and love. With their presence they have helped me to become a more highly evolved person and have opened a path for me to share this experience with others, the best job in the world! I feel truly honored to have for some reason been chosen by the dolphins and the whales to speak for them, protect them, and bring others to experience their mystical, magical, healing presence."

Chapter Six

Wild Birds

Bongo Marie

"Have I ever told you about Bongo Marie?" Raphaela asked innocently, as we sat one day drinking tea and enjoying the antics of her mini-Macaw, Dax. "No? I must. Bongo Marie was an African gray parrot who lived with Sally Blanchard, the publisher of *The Pet Bird Report*. Bongo Marie is quite famous in avian circles. Sally writes and talks about her frequently."

We had been discussing a bird called Halftone, an African gray who lives with two of my friends here in Eureka. My friends are printers, which explains Halftone's marvelously descriptive name. He is a charming bird, and I always enjoy seeing him, but nothing could have prepared me for what I was to hear about Bongo Marie. I begged Raphaela to go on.

"Well, Sally says that some African grays seem to think they are better than other parrots they live with. Bongo Marie definitely did. She always acted as if Paco, a double yellow-head who also lives with Sally, was not quite her equal. Bongo Marie liked to call out Paco's name over and over until he responded. Then Bongo Marie would say, 'Paco, be quiet!'

"One day Sally was making dinner. She had Bongo Marie's cage near the dining room table, and Paco's cage near the door. Sally bent down and took a Cornish game hen out of the oven. Bongo Marie climbed down the side of her cage and eyed Sally's dinner quizzically. Suddenly she threw up her head and in a frantic questioning voice exclaimed, 'Oh no! Paco!'

"When she stopped laughing, Sally explained to Bongo Marie that the bird on the platter was not Paco. 'Look, Bongo Marie,' she said. 'Paco is right over there.' The gray looked towards Paco and then said, 'Oh, no!' Then she laughed hysterically, as if to let Sally know she had been joking all the time."

Raphaela told me that this story is famous because Sally has printed it several times and often tells it in her seminars, but the anecdote also has another life as a bad example! A bird show trainer named Steve Martin uses Sally's story to demonstrate what he feels is the tendency of bird people to anthropomorphize their birds excessively. Blanchard writes in *The Pet Bird Report* that on his video Mr. Martin misquotes the story, and then says,

> *You know what? That whole story is fiction. It has to be fiction and I guarantee you it never happened that way. There is not a bird in the world that I know that has the capability of conversation or language like that. . . . Unfortunately it's put across by the quote unquote expert as fact. I think that's sad because it allows and encourages people to go home and interpret the behaviors that their bird does in such an anthropomorphic way that takes them farther and farther from a positive relationship with their bird.*

I found Mr. Martin's denigration of the Bongo Marie story fascinating, because Sally Blanchard is exactly the kind of knowledgeable, practical, dedicated person whose careful observations used to be the very backbone of natural studies. Rupert Sheldrake points out in his interesting book *Dogs That Know When Their Owners Are Coming Home* how extremely useful to science the work of people like Sally Blanchard is, or could be:

> *One of my favorite books in biology is* The Variation of Animals and Plants Under Domestication *by Charles*

Darwin, first published in 1868. It is full of information that Darwin collected from naturalists, explorers, colonial administrators, missionaries and others with whom he corresponded all over the world. . . .

Since the time of Darwin, science has increasingly cut itself off from the rich experience of people who are not professional scientists. There are still millions with practical experience of pigeons, dogs, cats, horses, parrots, bees and other animals, and of apple trees, roses, orchids and other plants. There are still tens of thousands of amateur naturalists. But scientific research is now almost entirely confined to universities and research institutes and carried out by professionals with Ph.D.s. This exclusivity has seriously impoverished modern biology.

Also in favor of the Bongo Marie story is the fact that no one but an African gray could have come up with it.

Did Bongo Marie know what she was saying? Blanchard writes,

I have no idea what went through Bongo Marie's incredible little brain to create the scenario with the Cornish game hen, but I can guarantee I quoted her accurately. . . .

Do grays use language? Yes, of course! They have their own vocabulary of sounds they use for verbal communication. Do they know what they are saying when they speak in our native tongue? Rudimentary language is simply an association between sounds and the objects, situations and companions they identify. In this sense, companion grays do not simply mimic. While much of what grays say may be simply mimicking or nonsense babbling, often they show themselves capable of understanding labels for many aspects of their lives. This is certainly a rudimentary sense of language. At this point in time, anyone who thinks otherwise is blind to the accurate observations of hundreds of people who are sharing their lives with these remarkable parrots.

Sally Blanchard's observations would have been good enough for Darwin. They are more than good enough for me. Darwin would have also approved of the more than three hundred readers who accepted an invitation from *Bird Talk* magazine to describe the intelligence and language skills of their African grays. An article in the November 1999 issue of the magazine gives some of the most amazing examples of, well, bird talk you could ever hope to see. Here's our favorite. A *Bird Talk* reader named David D. Scherr, of Missouri, wrote that an insect was flying around in the cage of his Congo, Rio. After watching the insect intently for a while, Rio said, "Is it a bird?"

You tell me that Rio wasn't talking!

Wildness

Do you think you would like to live with a wild animal? You can, and you don't have to raise a lion cub or bring a bear into your home, cither. All you have to do is bring home a parrot. Dogs, cats, and horses have been sharing their lives with people for many, many years. Recent DNA evidence suggests that the transformation of wild wolf to domesticated dog occurred over ten thousand years ago. Cats became domestic much more recently, about five thousand years ago, and in fact they are still notably independent and revert with relative ease to wild living. I often think they have never left it. Horses joined us as draft animals at about the same time as cats. The first record of a horse being ridden is about 35 hundred years ago.

No parrot, even one you hand raise yourself, is more than three or four generations away from at least 35 million years of wildness.

This means that living with a bird is fundamentally different from having a domestic animal companion. After so long together, dogs and cats are used to our ways, and we to theirs. Not so with birds. They are surprised by us, and equally, they are capable of amazing us with their sheer otherness.

Raphaela swears that there is nothing in the world as knock-down hilarious as a baby parrot. Her Yellow-collared Macaw, Dax, has about 50 toys. He picks up and plays with every one of them at least once a day. He talks, laughs, does party tricks, hangs upside down in the apple tree in her back yard, screams, flings his food about, turns somersaults, and entertains her nonstop.

There is also such a thing as bird adolescence, when the amusing antics become wilder. An adolescent bird will settle down again in due course, just like your teenager, but the papers are full of ads for "two-year-old Macaw, Bargain." We just mention that for your information.

When a wild parrot comes into a domestic environment, he or she will naturally use the same traits that have helped them live their lives in the wild to adapt to their new lives with us. Bonnie Monro Doane and Thomas Qualkinbush, the authors of *My Parrot, My Friend*, describe certain wild parrot traits that make them wonderful animal companions and happy dwellers in our home.

First and foremost, they are excellent colonizers. Parrots are flexible and adaptable, good at moving into new environments and exploiting them for their benefit. When farm land encroaches on their habitat, they are willing and able to eat the crops and drink from the livestock troughs. When they move into your home, they are equally willing and able to eat from your plate and drink from your cup.

Next, they are hierarchical. They are used to living in well-developed societies in which they dominate weaker birds and are submissive to stronger ones. This means that they can fit right into your domestic hierarchy, usually as king or queen. Size definitely matters. You are bigger, but if a parrot sits on your head or even on your shoulder, he may believe *he* is bigger. Actually, he may believe that anyway.

Another trait that makes parrots very suited to domestication is their strong natural urge towards pair-bonding. When a parrot moves in with you, that definitely works in your favor. You and your parrot become a pair. The parrot adores you, regurgitates food for you (thanks), preens your eyelashes and eyebrows, bows and ruffles up, and invites you to preen him. It sounds wonderful and it is. It can also get much too exclusive, and lead your parrot to become jealous of the human with whom you have also pair-bonded.

Stories of parrots who dislike their person's spouse abound. In fact, that was precisely what I had been telling Raphaela about my friends' African gray, Halftone. Halftone is strongly bonded with the husband, David. She regards David's wife Barbara as an interloper and not a very welcome one at that. When Barbara comes into the room Halftonc is apt to start yelling, "Barbara, you brat," or "Go to work, Barbara."

Clearly, living with a wild animal has its emotional perils. Halftone has the refreshing candor of an animal who hasn't learned the perfect diplomacy of a dog, by which the dog is able to convince each of its people that he and no one else is the dog's special favorite; nor yet the universal disdain of a cat, by which each person feels, at best, equally and barely tolerated.

Long centuries of domestication has brought us to an accommodation with dogs and cats that serves us both well. Will domestication turn out to be equally good for parrots? The jury is still out. For now, all we can say is that living with a parrot offers a unique chance to be with a wild animal in all its charm, and to witness its life in as much freedom as you are able to provide for it.

Be warned, though. Parrots are complex, demanding creatures. If you are the sort of person who enjoys stroking your cat for five minutes before going out the door to your next activity, don't even consider a parrot as a companion animal. Neglected birds—meaning birds you don't give a *lot* of time and attention—may pluck out their feathers, mutilate their flesh, and prolapse their cloacas in their misery and unhappiness. It's not pretty and you won't enjoy it.

Parrots need stimulation and a reliably high level of interaction with loving caretakers who are committed to their well-being. Also, be aware that your commitment is a long one. Parrots are very long-lived. You know that bumper sticker that says, "A puppy is for life, not for Christmas"? In the case of a parrot, that life may be 40, 50, or even 60 years.

At the same time, domestication may also offer parrots their only chance of survival. Their native rain forests are disappearing. A sanctuary with us may be their only hope of avoiding extinction.

Perhaps parrots are somehow aware of the lesson to be learned by comparing the fate of dogs and that of wolves. Those wolves who threw their lot in with humans, and became dogs, now exist by the hundreds of millions everywhere in the world. Those who remained wolves, by contrast, are extremely thin on the ground. The same is true for cats—there are many millions of domestic ones, and almost every wild cat is endangered or already extinct.

Wolves, at this moment, are being reintroduced to habitats that need them badly (Raphaela talks with wild wolves in Chapter 7). It may be that parrots, too, can be put back into their habitats in time, if only their habitats can be somehow secured for them. In the meantime, parrots are here. We may as well enjoy them for the priceless companions they are.

The Legacy of Breeze

Raphaela's love affair with parrots began with her seduction by a Black-headed Caique named Petruccio. The locale was a specialty pet store, and the reason for the visit was to find a bird that would not replace, because that would be impossible, but begin to fill the great hole left in several lives by the death of Breeze.

Breeze was a canary whom Raphaela took care of on weekends when her friend Donna, who stayed in Raphaela's home and did graduate work at the University in Berkeley during the week, returned to her home in Davis. Breeze had become the innocent victim in a long-running disagreement between Raphaela and her cat, Sophia.

Raphaela had a well-intentioned desire to improve Sophia's diet. Sophia had a deep-rooted conviction that dry crunchies were far superior to homemade organic cat food, and she should not be made to eat what she did not like. Raphaela had been offering Sophia healthy cat food, and Sophia had been refusing to eat, for several days when Raphaela returned home to find, to her horror, that Sophia had killed and eaten Breeze. When Raphaela expressed her shock and outrage to Sophia, the cat, also very angry, replied, "If you had given me anything decent to eat, I wouldn't have had to do it."

The anger and hurt on all sides were eventually healed, but Breeze was never forgotten. Eventually Raphaela and Donna made their way to a specialty bird store in Orinda to look for a new bird. There they encountered Petruccio. She was the poster bird for Caiques; to see her was to adore her. She came to the door of her cage, expecting to be let out. Of course they let her out, and she immediately ran up Donna's arm, played with her buttons, tweaked her earrings, explored her body, and finally climbed into Donna's pocket, as if to say, "Ready! Let's go!" (A comment on Petruccio's sex: did you think she was a male from her name? So did Jill, the store owner, at first. There are countless birds with boys' names who are girls, and vice versa—it's almost impossible to tell their sex without a DNA test.)

Petruccio was not for sale. The most animated and outgoing Caique in the world, Petruccio lived with Jill and was her mascot and chief public relations officer. However, at that moment Jill had no baby Caiques to offer. Donna and Raphaela found a breeder in El Cerrito. There Donna found Breeze's successor, a Caique she named Dominique Bramwell Kachinskas, Esquire, and nicknamed Nique for short. Raphaela was equally smitten. Nique's sister Sancho (see what I mean about the names?) went home with her and her husband Roger.

Nique and Sancho, Roger, Raphaela, and Donna became a tightly knit flock in the birds' eyes—so much so that it could be alarming to strangers. One day Raphaela was sitting in the kitchen talking with a visitor. They were joined by two pint-sized parrots in attack mode: pacing the kitchen counter, their eyes pinning, muttering ominously.

Raphaela assured the nervous visitor that the birds' wings were clipped and they could not get off the kitchen counter. The humans moved to the dining room. Then the visitor saw both of the parrots carefully inching their way down the electric cord that plugged in the toaster oven. "I'll call you," she called to Raphaela as she beat a rapid retreat. Two-ounce Nique had transformed himself into The Terminator.

When Roger and Raphaela separated, Sancho went with Roger, and Raphaela moved to Davis with Donna and her husband and son. Raphaela missed Sancho badly, so to make up for it she brought home a Yellow-collared Macaw whom she named Dax. Sancho still visits her brother Nique in Davis from time to time. Phoebe, a Dusky Pionus, has also joined the household—a gift from one of Raphaela's clients.

Dax, Nique, and Phoebe, along with Sancho when she comes visiting, make up a wild, energetic bird household. The four birds are cordial up to a point. At other times the social fabric frays. Raphaela was filing Dax's nails one day, a procedure that requires him to be wrapped in a towel and during which he screams bloody murder. Phoebe climbed onto the chair and hopped down Raphaela's arms, wings out, eyes pinning, as if to say. "Keep holding him down. I'll finish him off."

The Davis house is also home to an Irish Wolfhound named Kiernan, and two Whippets, Fl'ar and Fiona. This makes for some interesting interactions. There have been no further eatings, but the mixture of domestic predators and wild birds means the humans have to keep a pretty close eye on the birds.

Nique has a special relationship with Kiernan. They came into the house at about the same time, when Nique was a baby bird and Kiernan just a puppy. They grew up together, and Nique spent a lot of time keeping Kiernan company while Donna was away at work. Kiernan, now a full-grown Wolfhound who weighs 150 pounds and resembles a medium-sized horse, plays with two-ounce Nique and has been known to lick his beak affectionately. Nique, for his part, when he wants to go somewhere just hops on Kiernan's back. He has also been known to hop from the kitchen counter onto the back of one of Whippets. They go nuts.

In theory, Donna or Raphaela watch the birds whenever they are out of their cages, but with the best of intentions, vigilance can fail. One night the humans were having dinner when someone said, "Where's Nique?" Everyone leaped up from the table and rushed about the house looking for him. He was discovered in the living room, in the dark, sitting on Kiernan's back and saying, "Oooh! Oooh! Oooh!"

Raphaela asked Nique what he and Kiernan talk about when he goes for a ride. Nique said, "I think of him as a bridge. I want to go somewhere, and he's going. I just hop on. Sometimes I tell him where I want to go—to the seed stand in the living room! Or to the living room, to watch the wild birds outside. But sometimes I just go wherever he's going."

Kiernan put in his two cents, too. "When Nique first jumped on my back, I was so surprised! I took him down the stairs. Donna came rushing after us. It was really fun! I ran out into the garden. I thought Donna was playing, but then she grabbed Nique and took him back inside.

"Nique is special," Kiernan went on, "because I grew up with him. We kept each other company when Donna was at work. I know him best of all the birds. Also, Nique gives me food. He drops things on the floor and I eat them. Sometimes I lick his beak to get the wet food off. He's my friend."

Dax: I Eat What the Wild Birds Eat

When my life is too much the same all the time, I get bored. I like to go a lot of different places. I have different places in our house. I never stay in one place too long. I like to go into the food place, and then I go into the sleeping place and then I go where the dogs are. Sometimes I go where the cat is, but I don't stay there very long.

Raphaela takes me outside and there's a tree there and then she does something to make water come onto the tree. The water makes rainbows. I love that. Nique and I hang from our feet in the tree. We scream. I love screaming.

Flying is the best. I used to think I couldn't fly. Raphaela does something to my wings. She says it's because I'm so special and she wants me to stay with her all the time. I don't mind. She's good to me and there's always a lot of food. I love eating. Sometimes I go outside and I sit on her finger she runs and we pretend I'm flying. She screams, "Dax is flying!" She likes to scream, too.

Then I found out that I can fly after all. Not just pretend, on Raphaela's finger, but really. I was so surprised! I was outside in the back yard playing around in our tree, and I started moving my wings and pretending I was flying, and the next thing I knew, I *was* flying!

There is another tree on the other side of our house. It's much, much bigger than the tree where the water goes on us. I just kept flying up and up and the next thing I knew I was at the very top of that tree. I loved being up there. I could see Roger down below me and he was so little, he looked about as big as that dumb Phoebe who lives with us. I called out to him. Hallooo, Hallooo, I said. I said it a lot of times, and I laughed.

Then Raphaela came out of the house. She looked really small, too. She started calling me and telling me to come back into the house. I didn't want to. It was really fun to be outside. If I can fly, I can go anywhere I want to. No one can catch me and put me back in my cage.

I decided it would be fun to stay outside for a little while longer. It was starting to get dark. I never used to be able to stay outside when it was dark. I must be a really big bird now.

After a little while I spotted Roger. At first I couldn't figure out what he was doing, but then I saw that he had a ladder, just like the ladder I have in my cage. I like my ladder. I stand on it and play with my bell toys, but this ladder was too big. Roger was too big, too. He started climbing up into my tree and I didn't want him to. He looked scary to me, coming up my tree like that. I decided to fly away to another tree. This time I didn't call out, so no one knew where I went. After a while they all went back inside the house.

Then Roger came out of the house again and got on his motorcycle. I saw him start going up to all the houses, one after another. I laughed to myself because Roger thought I was inside one of those houses. I wasn't. I stayed right where I was in my new tree. When it got really dark outside, I found out it was just the same as inside, except there was more to do. There were more people and a lot more birds. I watched everything that happened, and I hardly went to sleep the whole night.

Then it got light again. I decided to fly to another tree. Just when I had decided on a nice branch to sit on, I started getting a lot of calls from Raphaela's friends. Raphaela knows how to talk to me in my mind, and a lot of her friends do, too. Her friends are kind of boring, though. They all said the same thing to me: "Dax, go to people. Go to a house. You aren't a wild bird." Shows what they know! I'm just as wild as any old bird, so ha ha.

Then Raphaela talked to me. She told me to come home. She said that she put my cage outside on the porch, and put big white sheets all around it so I would be able to find it. Ha ha! I laughed some more. It was funny, because I know where my house is. At least, I think I do. From up here they all look a little bit alike.

I don't care anyway because I'm not going to live in a house any more! I'm a wild bird now. Those people who keep calling me are always saying I'm not a wild bird, but I am. I go where the wild birds go. I'm soaring! I never want to stop flying. Back when I lived in my house, I used to step off the counter and glide down to the floor. That's all I could do, and I always thought it was fun. Whee! I slide down on the air. Now I step off my tree just like it was the counter, and I start to glide but then I move my wings and I go up instead of down. It feels so good. I'm flying! I'm soaring! When Raphaela calls me again, I'm going to tell her, You should learn to fly! It's the best thing anyone could ever do. I'm not afraid of anything any more. Even if the cat came I wouldn't care, because I can fly!

Oh, now there's Raphaela talking to me again. She wants to know if I'm hungry. Hungry? That's something to think about, because Raphaela always used to give me a lot of good things to eat. I like bananas, and apples, and I like pears and peaches and I like crackers and seeds and peanut butter and almonds. I'm sure if I just find the right tree, I can have anything to eat I want.

I'm just going to tell Raphaela I eat what the wild birds eat. Ha ha! I eat what the wild birds eat!

Someone just came out of that house over there and pointed at me. Now she's going back inside the house. Now Raphaela's talking to me again. She's saying that I should stay where I am because she's on her way to get me. I wonder if I should go home now. Oh! What was that noise? I think a car made it. I don't like that kind of noise. I'm going to fly away to another tree.

Maybe I am a little hungry. I think I would like to eat some peanut butter. I wonder where the wild birds get peanut butter. It's starting to get dark again. I'd better stay here in this tree. Maybe tomorrow I'll find something good to eat. Yes, I'm quite sure I will; and then I'll fly some more. Right now I'll just sleep. I'm feeling a little tired, and it's not as warm out here as it is in my house. I wonder which house is mine. I'll fly around in the morning and look for it.

I'm glad it's light again. I like it better when it's warm and light out. I'm going to fly to another tree. Hey! Look down there! A tree with food on it! I'm going to fly to that tree. I know what those things are. Those are plums! This is a good tree. I'm going to eat some plums. Good! Now I'd like to have some peanut butter. Maybe I should go inside that house. It's sort of like my house, but I don't think it is my house. There's someone coming out the door. It's a little person. Hello! Hello! Hello! Now he's going back in the house. Now someone else is coming out—a bigger person, more like Raphaela. She's holding out a piece of apple! I'd like to eat that. I think I'll get on her arm.

This is a nice house! It's not my house and I wish it was my house, but there are a lot of good things to eat here. I'm going to have some of this apple, and now I'm going to eat some of these seeds, and then I'm going to eat some of this banana.

Who's that? Someone is calling me. She's saying Dax! Dax! That's me! I hear her. It's her! It's Raphaela! I'm screaming! I'm flying to her! Raphaela! Raphaela! Hellooo! Hellooo! Let's go home!

"Thank you!" said Raphaela to the kind woman who had been feeding Dax a smorgasbord on her kitchen counter. "How did you find Dax? And how did you know to call me?"

"My son is the one to thank. He spotted your little bird in our plum tree, and he came and got me. I went out to look at the little guy, and then I held out a bit of apple and he came inside with me. We quickly closed all the windows and the door. Then I made up a "Found Bird" poster on my computer and sent my son down to the grocery store on the corner to put it up. He saw your "Lost Bird" poster on the bulletin board there, so he pedaled right home and we called you."

"Come on, Daxito," Raphaela said. She held out her finger, and I hopped on, even though there was still a lot of peanut butter left in the jar. "I eat what the wild birds eat," I thought. It was a great adventure. I went where the wild birds go! And now, I'm going home.

The Parrot Interviews

As it turned out, this wasn't the only time Dax flew away. It had happened once before, and this time Raphaela located Dax with telepathic communication. Dax showed her that he was beside an "enormous brick building." Raphaela naturally assumed he had somehow managed to fly all the way to downtown Berkeley, or even to the University campus. However, when she walked out into her own back yard she heard Dax faintly calling out his Hellooo, Hellooo, and found him crouched behind her brick chimney—an enormous brick building to minuscule Dax.

"How is it that Dax could fly away?" I asked Raphaela. "Surely you keep his wings clipped!"

"Of course I do," she said. "What I didn't realize is that the last groomer I took Dax to had given him what they call a vanity clip, where the first two flight feathers are not clipped. With his other feathers partially grown in, and a good stiff breeze, little Dax just took off. Sally Blanchard wrote a story in *The Pet Bird Report* about a Macaw of hers named Jobo who did the same thing. It took her three days to get Jobo back, using his mate, Bojo, as a decoy to entice him back to his cage."

"Has Nique ever flown away?"

"No, but Nique is a Caique, not a Macaw. Caiques are little flying rocks. The Amazonian Caiques—the Black-headed Caique and the White-bellied Caique—live on opposite sides of the Amazon, because neither can fly well enough to actually cross the river."

I told Raphaela I was impressed by Dax's flying prowess and interested in how much he loved flying.

"Yes, when I tuned in to Dax and he told me about it, it reminded me of one of those incredible flying dreams that come to us if we're lucky, and you wake up so disappointed that you can't really fly. Only, Dax was really flying! It must be the central experience of a bird's life. Researchers who study domestic parrots are finding that being allowed to fly is absolutely critical for a bird's well-being. Even though I was petrified for Dax the whole time, I can't feel really sorry he got to have that experience."

"So domestication might mean survival for parrots, but it might also mean losing their essential nature?"

"I don't really know. I'll tell you what—let's ask the birds."

Raphaela settled down to interview Dax, Nique, and Phoebe. I was invited to join in the telepathic link if I could. Dax had just had his shower (he loves to play in the shower with Raphaela), and was nodding nearby in his cage, so we started with him.

"Hi Daxie," Raphaela said telepathically. "We'd like to ask you about domestication. How do you like being a domestic bird?"

"It's good," Dax said. "Less work. I don't like to work. Macaws know how to have fun."

"Do you see any drawbacks?"

"There's not enough to do. I need more interaction!"

"Dax, I interact with you all day long."

"More! More!"

"What do you like best about your life here?"

"I like the Greenbelt. Flying is good. I like to go in the car. I like it when you have company. I especially like it when people learn to talk with animals. I help them."

"Yes you do, Dax. You love to chat with the people. You're a good bird. Let me ask you something else. Did you know when you came into this life that you were going to be a domestic bird? Did you choose that?"

"Kind of. I didn't know exactly where I would be in this life or whom I would be living with, but it's developing very well!"

I, in the meantime, was clinging to the telepathic link as best I could. When Raphaela asked him how he liked being domesticated, I heard Dax laugh and say, "She thinks she's domesticating me. Actually, I'm wilding her. The bad part is I can't fly as much as I want to, but we're living with people so we have to walk around like they do. They're a lot bigger than we are, but so what? What's the point of being big, anyway? You have a good life, though. The food is great, and it's a lot less work! You should be more like us. You should scream more. Screaming is good. You have some bad animals [I think he meant Sophia, the cat] but they are little bad animals, not big bad animals" [at least Sophia's not a jaguar]. "And you don't have any bad birds."

"What does Dax mean by bad birds?" I asked Raphaela.

"Little birds like Dax would be preyed on in the wild by big raptors like hawks and condors. Whenever I have the birds outside in the apple tree, they're very aware if a shadow crosses the tree. They all get nervous and start looking around for shelter. What did you hear Dax say when I asked him if he knew what he was getting in for in this lifetime?"

"I thought he said he didn't know. I got a sense that birds aren't into service lifetimes, the way a cat or a dog might be. They're just not that way. I think he said that when he found out you know how to talk, he laughed."

"That sounds like Dax, all right. Let's interview Phoebe now."

Phoebe, the Dusky Pionus, gives a much quieter impression than the flamboyant and dramatic Dax. She is gray and looks like a demure little Quaker quail. Her colors seem totally muted, until the light falls on them, when you see that her tail feathers are iridescent cobalt and magenta and orange. (Never underestimate Quakers.)

When Phoebe first came to live with Dax and Nique she appeared rather shocked by their ways. They love bathing and delight in joining people in the shower, but Phoebe refused to bathe, even in her bowl. The only thing that would encourage her was the sound of the vacuum cleaner, which Raphaela thinks reminds her of the sound of a waterfall and stimulates her desire to bathe. When all three birds go outside to play in the apple tree, and Raphaela puts the hose on fine mist for them to enjoy, Dax and Nique go crazy. They hang upside down, scream, flap their wings and generally live it up. Phoebe at first would just look at them askance. Finally she has begun to loosen up. She flaps her wings slowly, sedately, and with great dignity.

"Phoebe," Raphaela said, "What do you think about being domestic? Do you like living with us?"

"I do like it," Phoebe said. "I am well suited to it, as long as there is enough to do. I don't need as much to do as Dax. I would like there to be another Pionus. A boy," she clarified.

"Have you ever lived with people before?"

"Yes, in two previous lives. I liked those lives. To me, living with people is an experiment. I'm seeing how it works out."

"Did you make a choice to be domestic this lifetime?"

"Not so clearly. I thought I would be in the trees more. But this is all right. I've gotten much less afraid. There is a scary cat here. At first every time I saw her I screamed loudly. I wanted to warn the others, but then I saw that they weren't so afraid of her. Now I just tell the cat to go away. Sometimes I even chase her."

I tried to listen in to Phoebe also, and heard her say, "You want to know if I like being domestic, but you know what? This place is plenty wild enough for me. I'm more of an observing bird; I don't have to do all the time. But I like to help. This is a place where I can help, and gradually I will learn to do this. I can talk to quiet people if they come here. I bring a lot of peace. I can't talk to you any more because I am talking with Raphaela now. Dax can talk to two people at once, but I can't. Goodbye."

Then it was Nique's turn. Tiny little Nique is a very, very active bird. When we talked to him he was in his big cage in the living room, making the sound of the smoke detector. Raphaela asked him how he liked being domesticated, and he said, "It's good! But being wild is good too! Here there is lots of good food. I love to discover new food. I love my people. You might not think so because lately I've been biting them a lot, but it's just hormones, you know. I'm better now."

Raphaela told me that Donna hides new food in places where Nique can find it, which is a special joy for him. She confirmed that he has turned more aggressive since entering adolescence. She asked Nique if he had any other thoughts on domestication.

"No, not really. I'm afraid but then I get better. I like the big window because I can watch the wild birds out in the apple tree."

"Did you know you would be domestic in this lifetime?"

"No, but I'm glad."

I tuned in to Nique too and found that, like Dax, he could carry on a conversation with me while he was talking to Raphaela. He said, "Domestication? I don't even know what you mean. It's the same thing—we just go wherever we can. Right now we're here. Another time we might be somewhere else. I'm a pop-up bird." Puzzlement from me. "Yes, all of us are pop-ups. We pop up. A big bad bird comes and we hide, and then we pop up. We always do that. I am a religious bird. I am always praying. I am a thought of God. Yes, birds are God's messengers. He sends us to look and tell. Just the way we fly brings a message, if you know how to look at it.

"We worship the First Bird. The First Bird carried the First Seed, did you know that? You know that tree outside the window? That tree was planted by a bird."

When I told this to Raphaela she commented that it sometimes drives people crazy that their bird only eats a fifth or so of a piece of fruit, and flings the rest on the ground. But this is part of their role, the way they scatter seeds and plant new fruit trees. The volunteer apple tree in the yard *was* probably planted by a bird.

"Ask him whether he really said he was a religious bird. I might have gotten that wrong," I said.

Raphaela tuned back in to Nique, who said, "Yes, I worship and pray all the time. I don't make any difference between time for prayer and time for playing and eating and having fun. We are all here by the grace of the First Bird."

Whether he's domestic or wild, it seems that blessed little Nique has never left the Garden.

Petruccio: My Homecoming

I like people and they like me. There are exceptions, of course. I didn't like the people who stole me. In fact, if you want to know the truth, I hated them. I would have enjoyed killing them, only birds can't kill people very easily, because they're a lot bigger than we are. I think a jaguar, like the kind we have at home, might be able to kill them, but I didn't have a jaguar with me, so they got away.

Most people are pretty nice, though. Jill and I always have a good time with the people who come into her place. That's where we live. I sleep there at night, and Jill comes over almost every day to play with me. She calls her place a store but I call it our tree because that's where we spend most of our time. There's food there, and trees usually have food on them. There are a lot of birds who live there, too, and trees always have birds hanging around in them, so I think I'm justified in calling Jill's place her tree. To me, it just sounds better.

I'm a Black-headed Caique, and I'm a pretty special bird. Whenever people come to our tree, I always say hello right away. That makes them laugh, and they usually come over to where I stay and open up the door of my place, and then I go out and walk around on them. Sometimes if they have an opening in their clothes, I go into it. Sometimes they have little bags with them, and I might go into the bag. I just do it to be funny. I believe in always trying to get people to laugh, because they are a lot more fun when they laugh.

The people who stole me didn't laugh at all. That's how I knew something was wrong. They came into the store but they didn't say "hello" to Jill or "hello" to me. That's what people usually say, but these people, there were two of them, looked more like birds that are trying to sneak fruit from a tree that doesn't belong to them.

I put up the alarm so Jill would know there was somebody in the store I didn't like, but she didn't pay attention to me. Maybe she didn't hear me, I don't know. Anyway Jill was busy with some other people and she just called out, "It's okay, Petruccio." I still didn't think it was okay, but Jill said it was so I went with that.

One of the people I didn't like came over to my cage and put his hand right in. He started talking to me, and after a while I hopped onto his arm. Just habit, I guess. Then I hopped down his arm and got into his pocket.

I honestly can't tell you what happened after that. All I know is that I couldn't get back out, and then everything got really strange and frightening. I could tell we were leaving the store but

I couldn't see anything because I was inside his pocket. I thought Jill would come running after me, but now I know that she didn't know I was in the person's pocket. Otherwise, she would have. I know that because of how hard she tried to get me back.

The next thing that happened was that I was taken out of the pocket, not too gently either, and put in a room. I didn't have a special place to be, and there weren't any toys or bells or anything. So you can see that it wasn't a nice place, like Jill's. The worst part was that it was dark all the time. The other worst part was that there was nothing to eat except for seeds. Seeds are okay, but I need lots of different things to eat, especially fruit, and I didn't get any. I felt hungry and sad and scared. That part lasted for a long time—Jill says it was three days but I can tell you that it was a lot longer, or at least it seemed like it.

Then my life improved. I went in a car, and ended up living with a woman who loved me. She bought me from the bad people, but she wasn't bad. She was nice, but she wasn't Jill. I missed Jill a lot, but I couldn't figure out how to get back to her. I knew I'd have to just wait until Jill could find me.

The thing that made me feel better was that after a little while, Raphaela talked to me in my mind. I know Raphaela pretty well, because she came to Jill's store and talked to Jill's friends about how to talk to birds. Sometimes she'd talk to me, just to show people how it works. I would talk to anyone who wanted to talk to me, and Raphaela always said I was a good bird and a big help to her.

When Raphaela talked to me, I showed her a picture of the woman I was staying with. I told her that I was all right but I missed Jill, and she told me that Jill was looking for me and promised me that she would find me, no matter how long it took.

I stayed with that woman a long time. Jill said it was eight months. It seemed like forever, but I have a good memory and I'm glad to say that Jill does, too. Three or four more times, Raphaela talked to me and told me that Jill was still looking for me. That made me feel better. Raphaela told me that every single month, Jill sent my picture to all the vets she knows. Not just bird vets, but all kinds of vets. I have something wrong with my beak, an "abnormality," and I have to go to the vet once in a while to have it filed down. Jill knew the woman who had me would have to take me to a vet some time, and then she would find out where I was.

You know what? It worked! The woman did take me to the vet, and the vet knew who I am! I heard it all, so I know. The vet had a picture of me from Jill, and she said, "I think this is the bird in the picture!" Then they held up the picture next to me, and I heard them saying, "Yes! It's her! This must be Petruccio!" I squawked and hopped so she would know it was me.

I don't know exactly how the vet got the woman to leave me at her office, but somehow she did. Then the next thing I knew, Jill was there! When I saw her I screamed and jumped, and Jill picked me up, and we were so happy! We both screamed and jumped for a long time! If we could have flown, we would have, and I mean Jill, too, not just me.

Then I went home, back to my place, my tree. Jill kept saying how happy she was to have me back, and not just because I was "valuable." She said bad people had stolen me, but now I was back and she was going to make sure I was never stolen again. That's good because I want to stay right here. My tree is the best tree and I never want to leave it again.

The Wrecking Crew

"Under Siege," proclaimed the headline of an unusual article in the Sept 27, 1999, issue of *People* magazine. The besieged in question are the inhabitants of a little town called Pine Mountain Club, California, about 90 miles north of Los Angeles. The besiegers are a gang of 20 adolescent California condors who were released into the wild after being bred in captivity in zoos. The townspeople call the condors the "wrecking crew." Since coming to Pine Mountain Club in August, they have had quite an impact. They've taken chunks out of patio furniture, stripped insulation from pipes, torn up roofs, and eaten a mattress.

This was amazing. I thought back to a class I'd taken years before through UCLA Extension at the Los Angeles Zoo. The highlight of the class was a chance to look at the Zoo's two California condors. I remembered them vividly—gigantic creatures (we were told they had a wing span of nine feet), black all over with bright red-orange heads, their heads and feet quite bare of feathers in the way of scavengers, who must put their heads into the bodies of their dead prey.

They sat looking absolutely prehistoric on a high perch in a lofty aviary. I was enormously impressed. They were two of just a handful of condors in existence. Now I read in the *People* article that condors had been down to their last six individuals when the breeding program began. Six! Talk about the brink of extinction! The two I saw, then, may well have been the ancestors of the Wrecking Crew. It was thrilling to think that there were now at least 20 of them, and that they were out there in the wild, eating mattresses.

But wait—mattresses? There was something wrong with this picture. I picked up the phone and called Raphacla. After reading the article to her I asked if we could talk to the condors. Raphaela agreed at once. She has a special interest in raptors, awakened when her late dog Petey was reincarnated as hawk. We decided to try and find out what the story was on the mattresses, and more generally to ask the condors how they were faring in the wild, what they felt were the pros, and cons if any, of wild living, and what their relations were with the people of Pine Mountain Club.

Raphaela asked me if I wanted to try listening in on her telepathic link. Of course I did. So when she settled down to call the condors, I did too.

I ask for nothing better from life than these moments when Raphaela and I put down the phone, quiet our minds and call out to the animal we are trying to reach. It's a moment when magic, real magic, hovers and then alights. Like what happens when my string quartet makes contact with the spirit of Beethoven, it's a moment my soul needs, the way my body needs air.

This was my third time to participate in an interview since Raphaela and I had begun working together. The first was the interview with Yabis, the gentle gray whale who had been hunted by the Makah tribe. Next had come my solo talk with Matrix, the spirited horse who had once been part of Raphaela's animal family. Now I was trying to reach a condor, and had every reason to think I would be able to. But *how was it happening?*

Whatever the explanation, the fact that telepathy works is profound and life-changing. There exists some mechanism in the universe that allows two women, linked by modern telephony but physically about three hundred miles apart, to reach out with their minds and contact the mind of an animal. If you believe, as I do, that it happens, and you don't want to negate the entire rest of your understanding of the way the world works, you would naturally like to find some comforting, at least remotely plausible explanation.

Rupert Sheldrake (in *Dogs That Know When Their Owners Are Coming Home* and also in his book *Ten Things You Can Do to Change the World*) suggests a possible explanation. His idea is that there exists a "morphic field" of telepathic connection between people and the animals with whom they have close ties. This field can be somehow stretched over great distances, and can be used not just for communication but also for guiding lost or left-behind animals back to their people even over great distances.

It is brave of Sheldrake, who thinks of himself as a scientist, to proffer this thought, and the explanation strikes me as a good start, but I can't feel it is broad enough. Sheldrake is interested mainly in telepathy between people and their close animal companions, but the telepathic communication Raphaela and other communicators do takes in a much wider field.

Take this situation, for instance. It's hard to see how there could be a morphic field built up of emotional connection between us and the condors. We don't know these condors; we don't know *any* condors. We don't have an emotional relationship with them. Raphaela is sympathetic to them, as she is to all animals, and feels positively towards raptors generally, but she had never so much as seen a condor, and she still hasn't. I saw that pair in the Los Angeles Zoo years ago, but to be honest, we weren't that close.

I think that what makes telepathic communication possible is both simpler and more wide-ranging than Sheldrake's morphic field. I think it works because we're all one being. It only seems like we're "connecting" to another being telepathically; in fact, we're just accessing another part of ourselves, or, let's say, another part of our own consciousness.

When I was a kid I used to lie in bed at night, speculating about consciousness. The point that interested me was, how was it that I, a big kid in the second grade, could remember things that happened to that *infant,* me in kindergarten? What was it that connected my earlier self, so little and dumb, with the mature, smart person I was now?

I decided it had to be the inner sense that made me think of myself as "I." It had traveled with me throughout my life, and had been constant through all my changes. But, I reflected, it can't just be something that belongs to me, because all the other people around me also think of themselves as "I," a fact I had gleaned through conversation with my mother and my brothers. They thought they were "I." I thought so, too, so it must be the sense of "I" that we all had in common.

This was how I reasoned as a child. Now, looking back, I think I was onto something. What I used to call the sense of "I," and that I now call consciousness, is the link among all apparently separate beings, uniting people to animals as well as to each other. To a condor, he's "I." He, and I, and everyone else, shares the same deep, constant level of consciousness. Why shouldn't we use that underlying level of community to talk to each other, regardless of space, time, or species?

Having cleared that up for myself, I felt better, and renewed my call to "a condor in or near Pine Mountain Club, California." And sure enough, there he was!

Or rather, there they were—I had the feeling that I had connected with not one condor but several together. My mind filled with an exhilarating sense of wildness and freedom. They seemed wonderfully adolescent—confident, scornful. The first message I got was, "We don't like you so much. We've been watching you, and we don't like what we see. You people think you're so great, but you aren't. We don't see you flying. Where do you get your meat?"

No answer occurred to me to this excellent question, so I proceeded to the first of the topics Raphaela and I had agreed upon: How are you doing in the wild? Do you like it?

The response that came back was, "Flying! Flying! Flying! Flying!" It wasn't the word, but a sense of how truly marvelous it was to them to stretch out their giant wings and just go. Then they said, "Eating! Eating! Eating is good. We are eating!" Then they laughed. Ha ha! "We do what we want," they exulted. "This is being in the wild!"

I laughed aloud with them. These condors were entirely exhilarating presences. However, I had work to do: I glanced down at the paper where I had jotted the topics, and picked up the next one. Do you know how close you came to extinction? "You were down to six condors! How do you feel about that? Do you care?"

"Hey, that's right," they came back. "We're glad you asked that. We used to own all the land around here! What happened?"

I knew, in a general way—it was the same old story of humanity moving in, taking over, the environmental degradation, the slide to extinction. However, I despaired of explaining it to these kids, and anyway, it didn't seem like the answer they were looking for. I just waited for him to go on.

"Okay," he said, "it doesn't matter because the land is coming back to us. We're getting it again. We're the watchers. We watch the land, and we clean it. We'll make the land clean again, like it used to be. That's why we're so strong—so we can eat, and clean."

"Elizabeth?" Raphaela's voice came from the phone. "Did you get through?"

"Yes," I said, still marveling at the wildness I had just encountered. "Did you? What did you get?"

"I did, too." she said. "They're huge! I had no idea! They showed me where they are; I saw a feeding platform that the researchers maintain for them, to help them as they're getting used to living on their own. Do they have those, do you know?"

It didn't say anything in the *People* article, but I later verified from the Los Angeles Zoo Web site that yes, there are platforms maintained by scientists in the reintroduction zones from which they assist the released condors in various ways. Condors have even been trapped and moved to safety when they appeared threatened, for example during wildfires at Big Sur in the summer of 1999.

"What else did you get? Were you speaking to just one condor? I think I had a gang of them."

"Just one, I think, perhaps a slightly older bird, but he seemed to be with some others. He said it was cold where he was. He showed me that he was at the top of a very tall pine tree. He said they have very, very good eyes. They see everything from mountain goats, which he tells me are rare; to little rodents. He's flooding me with visual images. It's fascinating."

"Did you ask him our questions?"

"I asked him what he thought about people, and he seemed interested in us, but cautious. He says they like to watch us just as much as we like to watch them. He sees people watching him all the time! Then I asked him what he thought about being in the wild. He says they miss the food they used to get in captivity, but—these are his words—'the memory is fading.' The platform he's showing me is very spacious, and there are bits of food clinging to it. They might still be using them even for food they find themselves.

"Then I asked him about condors being down to their last six. It didn't seem to faze him. He just told me that they have lots of vitality. He said, 'Give us half a chance and we'll make it.'"

"Isn't it amazing to talk to someone so wild?"

"Yes, I'm awestruck. Wait, I want to ask him one more thing: what does he want to tell people? Does he have any message for us?" Raphaela put down the phone again, and while I waited I couldn't resist putting in my own similar question. Did they have any message for me?

"Yes," came the answer, "we do. We'd like to come visit you. Where are you? And do you have meat?"

My immediate sense of alarm had nothing to do with the fact that Pine Mountain Club is five hundred miles south of Eureka. Although, as I thought about it, I remembered that condors are quite prodigious fliers who think nothing of flying a hundred miles for lunch. Whether it was realistic for them to visit me or not, my instinct was to run for cover. Sure, I thought to myself. I'm going to tell this condor where I am so he can come eat my dogs or my mattress.

Then I calmed down and realized that I was reacting to the fine sense of wild menace from the great bird. It really was exhilarating, the very thought of seeing a bird with nine-foot wings in the air above me. Scary, but exhilarating. I still didn't tell them where I was.

Raphaela returned to the line a few moments later. "He says we should just leave them alone. They'll be fine."

"Leave them alone," I echoed. "The same thing Granny says." The words that Granny, the wise old orca Raphaela and Mary Getten had interviewed, came back to us at the same time.

When we had asked Granny what she most wanted to tell humans, she had said: "Leavc us alone. We do not need to be managed."

I sighed, knowing that with the best will in the world, it was not so easy. With orcas, yes; we need to pull back our presence in the ocean, leave more salmon for the orcas, and watch them only from a respectful distance. With condors it was different. The reintroduction of a species that was nearly extinct requires a lot of intervention on our parts. In another sense, though, the condor's words were exactly right. Any intervention we do from now on should be in the direction of preserving the environment the condors need. In that sense, we could leave him alone.

"I get such a sense of mastery from him," Raphaela said. "Of ownership, really, like he does own the land. You know what? I think they can make it!"

I believe the same. They have already pulled back from the brink of extinction. There are 96 of them now living in captivity, being readied for release. Fifty-four of them are out there, freely flying.

Condors in the sky! There's hope for life, after all.

Chapter Seven

Los Lobos

The Hawk's Nest Pack

On March 29, 1998, three families of Mexican gray wolves—eleven wolves in all—vanished like smoke into the wilds of the Apache National Forest in southeastern Arizona. The wolves had been staying since January of the same year in acclimation pens—outdoor enclosures where they could be fed (on road-killed deer) and protected while they adjusted to local conditions. When the great day arrived and the pens were opened, all the wolves chose freedom immediately, without a moment's hesitation.

If there had been a question in anyone's mind—would the wolves perhaps choose to stay around and continue enjoying the comforts of an ample food supply?—it was answered at that moment. These 11 became the first Mexican wolves to roam freely in the southwest for at least 50 years.

Six of the wolves, a group known as the Hawk's Nest pack, were a family. There was the alpha wolf, a female known by the name 127; her mate, called 131; and their four pups. (Scientists discourage the naming of wolves who are to be released into the wild, but numbers seem to be acceptable, for some reason.)

The Hawk's Nest pack had been brought to Arizona from Wolf Haven International, a wolf sanctuary in Tenino, Washington. Wolf Haven is one of three sites where Mexican wolves are being bred in captivity and prepared for release—the others are the Sevilleta Wolf Management Facility in Socorro, New Mexico, and Ted Turner's Ladder Ranch, near Caballo, New Mexico.

Raphaela had been following the progress of the breeding program at Wolf Haven for some time. Like other wolf lovers around the world, she took a strong interest the adventures of the Hawk's Nest pack as they began making a new life for themselves.

Within a month there was bad news, good news, and more bad news. Just before the release, on March 26, 1998, a group opposed to wolf reintroduction (mainly cattle ranchers and farmers) had filed a lawsuit in the United States District Court in New Mexico. The lawsuit alleged that the Secretary of the Interior and the Director of the U.S. Fish and Wildlife Service had violated the National Environmental Policy Act in attempting to reintroduce wolves into the Blue Range Wolf Recovery Area (a larger area of which the Apache National Forest is part). This action did not stop the release, however, because the papers were not received until after the wolves had already taken off.

Good news came on April 21: the Hawk's Nest pack successfully hunted an elk. There was great rejoicing at Wolf Haven, since "their" wolves were the first with this major success. All the released wolves were from families of wolves who had endured at least five generations in captivity. There was much worry about whether their hunting instinct would still be strong enough; the Hawk's Nest pack showed it certainly was.

Then came more bad news. The male from another pair of wolves, the Turkey Creek pair, was shot and killed by a man camping with his family, who claimed the wolf was harassing his dog.

The events of the first month of Mexican wolf recovery neatly foreshadowed the whole of the next two years. Good news came from the wolves. They took to freedom perfectly, thrillingly, as if the five generations of captivity had been only a bad dream. The bad news was provided, as usual, by us. Lawsuits and shootings were ongoing. By the late fall of 1998, three of the four pups of the Hawk's Nest pack had been shot. The mother, 127, had simply disappeared. She lost her radio collar in August, and was last seen on September 19. The fourth Hawk's Nest pup, a female, was recaptured and returned to the Sevilleta breeding facility, with hopes of being released again later.

Of the original six Wolf Haven wolves, only the Hawk's Nest male, 131, remained in the wild. He was now a Lone Wolf and in need of a mate. He was recaptured in November 1998 and placed in another acclimation pen with a new female, this one from Ladder Ranch.

131 and his new mate, 486, were released again in January of 1999, but were observed traveling separately. This was not the idea, so later in January they were captured yet again and put back into the pen, in hopes of rekindling their interest in one another. Field notes from the Department of Fish and Wildlife's Web site state laconically that "the plan seems to be working." In mid-May, again to general rejoicing, three pups were born in the acclimation pen. The Hawk's Nest pack was reborn.

On June 6, 1999, 131, with his new family, was re-rereleased in the Blue Range Wilderness of eastern Arizona. The last report in the field notes, written in December 1999, says that they were "within their range in the Campbell Blue and Beaver Creek area, feeding on native prey." At least two of the pups were still traveling with the pack.

At this moment, at the end of the century, at the end of the millennium, the Hawk's Nest pack is alive and living in freedom. Not only that, they have company in the wild. A total of 34 Mexican wolves were released in 1998 and 1999. Five were shot (including the three Hawk's Nest pups), one disappeared (the Hawk's Nest female), one was hit by a car, and five were recaptured for later release. That means that at least 22 Mexican gray wolves are out there roaming the Arizona wilderness. There may even be more, since it is possible that some of the other wolf families have increased.

To think of them out there is absolutely thrilling. True, they face long odds. The Arizona wilderness is not a rich, wet, welcoming environment, like Yellowstone or southern Idaho, where gray wolves have been reintroduced quite successfully. It is very hot, with temperatures over 120 in the summer and scanty rainfall concentrated in just a few months. The land has been degraded by overgrazing and compacted by the hooves of cattle.

Yet the land does still hold herds of elk, white-tailed deer, and mule deer, animals with whom Mexican wolves have always lived in mutual interdependence. Wolves need them for food, and they need wolves to keep their herds healthy and vital and their numbers at a sustainable level. The odds may be long, but the wolves have a real shot at survival.

Then there is us. Five wolves shot? Out of only 34? What will happen when they begin to repopulate the area in serious numbers? What if the wolves prefer to prey on calves rather than elk? Will the ranchers and farmers accept them? Will the legal system continue to support the wolves, as it does today, or will the pendulum once more swing away from them? Of course, these are not questions an animal communicator like Raphaela can answer.

But she can talk to the wolves! "I want to talk to 131," Raphaela told me. "He sounds so amazing, and of all the Mexican wolves out there today, he's seen the most. He's a survivor who lost his entire first family. He was in the very first group to be released, and the first to be observed in a successful hunt. And now he's out there with a new family. I'd love to talk to him."

Raphaela invited me to take part in the interview, which we set for a morning in late December of 1999. It was to be one of the most stirring and enlightening conversations I'd ever had.

Peter, Ann, and the Wolf

I already knew something of the strong emotions wolves arouse. In 1994 an orchestra I played in, the Eureka Symphony, had a strange wolf adventure. It began at the end of our 1993 season, with an announcement that we would be presenting Sergei Prokofiev's *Peter and the Wolf* at an upcoming children's concert.

"I won't be playing with the Symphony anymore. This is my last concert," Ann Antonville, the principal cellist and my stand partner, whispered to me as we gathered up our music at the end of the spring concert. I was sorry to hear it. Ann was a fine cellist and an asset to the orchestra. However, she lived way out in the country, often complained about her long drive into town for rehearsals, and was planning to start home-schooling her daughter the next year. She seemed to have plenty of reasons to resign from the Symphony, and I didn't connect her decision with our upcoming performance of Prokofiev's chestnut. Wrong!

We had just started rehearsals for the children's concert when we heard that a small article about Ann had appeared in *Eco-News*, a local free tabloid with a strong environmental slant. The article reported that Ann Antonville had resigned from the Eureka Symphony in order to protest our programming *Peter and the Wolf.*

It appeared that Ann was a wolf activist, and felt that the piece made people hate and fear wolves. She did not condone the killing of the wolf, and she urged readers to show their support for wolves by boycotting the Eureka Symphony.

It seemed clever of Ann to use her resignation from the orchestra to get publicity for a cause she believed in, but I did wonder whether she had listened to *Peter and the Wolf* recently.

Of course, the wolf is not killed. He is captured by Peter, with the assistance of the bird and the cat. At the end of the piece, Peter and Grandfather take the wolf to the zoo, possibly saving him from hunters, who appear to be closing in on him in the forest.

This last was a prescient bit of programming, it now occurred to me. Wolves who survived in zoos are the very ones who are now being reintroduced to the forests where they have been hunted to extinction. I would have thought a wolf activist might even approve. A duck activist might feel outrage, as the duck does pay the ultimate penalty and is last heard quacking from within the wolf's belly. But for the wolf, at least, the outcome is as good as could be expected.

The next development took me by surprise. The article from obscure little *Eco-News* was taken up by the wire services as a horrid example of extreme political correctness. Suddenly our local flap was national news. Ann was ridiculed. Rush Limbaugh, disconcertingly, praised the Eureka Symphony for its brave stand against eco-terrorism. (We protested we'd done nothing that could be construed as putting us on the same side as him.)

Then we heard that *CBS Evening News* was coming to interview the orchestra. This seemed unlikely, but at the next rehearsal we found a camera crew in position and microphones pointed in our direction. Ann had already been interviewed elsewhere; now we were invited to say our piece.

Kay Pollard, a violinist, led off by pointing out that the wolf was merely a symbol for "all that was unknown and frightening in nature." Mike Ruud, a bass player, said that Peter's wolf at least wasn't as bad as Little Red Riding Hood's, who was portrayed as a murderous, thieving cross-dresser.

Then it was my turn. Wishing I had spent more time thinking through my position, and realizing that I had five seconds or less to make any point I might come up with, I told the camera person that I liked Ann and respected her love of wolves, but I thought she was mistaken. *Peter and the Wolf* does not make people hate wolves, it makes them love wolves, because Prokofiev has given the wolf such wonderful music. The wolf theme, played on four horns, is wild and haunting, and invokes all the mystery and beauty of the forest, just as wolves do.

Unwisely, I chose to reinforce my point on camera by attempting to sing the theme. The sheer ridiculousness of this action guaranteed it would make the cut, and I later had the privilege of watching myself bleat on national TV. Thus I squandered my possible 15 minutes of fame. More sadly for me, I forfeited Ann's friendship—she has not spoken to me since.

Despite my errors of judgment, I do still agree with myself. Wolves have undoubtedly been hated and feared by humanity, but their bad rap can't be laid at Prokofiev's door. I would be willing to bet that more wolf activists than wolf-haters have listened to and enjoyed *Peter and the Wolf*. The wolf is not so much villain as protagonist—the worthy opponent who brings out the best in Peter. Who knows, it may even have done something to repair the damage to wolves' image, after the Little Red Riding Hood PR debacle.

Victor Corbett, a friend in the viola section, wrote a poem that he read at the (unusually well-attended) Eureka Symphony performance. I thought it summed up all the ambiguities rather well. He has given me permission to use it here, with some slight adaptation.

Introduction to Sergei Prokofiev's Peter and the Wolf

(For the Eureka Symphony, 1994)

The wolf is a mammal much maligned
And vilified by humankind.
When a villain was needed by Brothers Grimm
They automatically turned to him.

He was portrayed blowing pigs' houses down,
Swallowing Granny and wearing her gown.
And not only authors showered abuse—heck,
It even extends to composers of music.

Like others who didn't treat wolves in a fair way,
Slander exists in the music of Sergei.
But let *us* be fair: this is not such a mystery—
Merely a slice of human-wolf history.

In days of yore, we did deplore
All wolves because there were so many.
Look around now, and you'll find, I vow,
Very few, if you run into any.

In truth, wolves have virtues we might emulate.
They're social and smart, they stand by their mate,
They kill just to fill themselves and their young—
And they have other virtues largely unsung.

So as you hear this popular story,
Which a listening duck might think of as gory,
Remember that nothing is all black and white
And the wolf only *plays* the villain tonight.

—Victor M. Corbett

This seemed fair to me, but when *CBS Evening News* played the story on TV a few days later, it had a different impact on me. They opened the segment with a piece of stock footage of wolves. It was an aerial shot, showing four wolves running about in the snow, as if trying to avoid some threat. The sound of a helicopter is audible in the background. Although nothing bad happens to these wolves on camera, you know in your gut that *something could.* The wolves aren't frightening, *we're* frightening. The wolves are absolutely magnificent.

Yes, Ann's protest may have had a whiff of late-twentieth–century political correctness about it. Yes, she might have chosen better targets than Prokofiev and the Eureka Symphony. But in another sense she was perfectly right. If she felt that we were in any way harming wolves, she had to protest. Anything we can do to protect the wolves, we should do. We have to have wolves.

Wolf 131

On the morning Raphaela and I had set for our first conversation with 131, I detected a touch of uneasiness in her voice. I could understand it—I felt it too. There is something momentous and frightening about the idea of talking to a wolf. A wolf! I found my mental image shifting back and forth between two pictures: one of the beautiful animals I'd seen on TV, the other an image of a yellow-eyed, slavering animal, compounded out of who knows what childhood tales and fears.

"I tried contacting 131 myself last night," I confessed to Raphaela.

"Did you? I tried this morning. Did you get anything?"

"A little. It was hard."

"That's what I found too. He's elusive."

"Well, he's supposed to be." Wolves that are raised at Wolf Haven or the other breeding facilities for the reintroduction program are given the very minimum exposure to humans. The idea is to make them wary of us, to preserve their wildness. We want them to stay away from livestock and, especially, away from areas inhabited by humans. The fear is for the wolves, who are in grave danger from ranchers if they approach their herds. Fear of humans is a very basic survival strategy for a wolf in the wild.

I asked Raphaela what she'd gotten from 131 on her first try. She said the wolf had indicated that he might be able to show her some pictures of his life, but that he wouldn't be forthcoming with opinions or philosophy.

"I got that he's not very introspective," she said. "He's very much in the present. He has to work. Making a living is his priority right now, as it should be. He has a family, after all."

I told her about my attempt to connect with him the night before. "I meditated for a few minutes and then I asked for the wolf called 131, in the Blue Range Wolf Recovery Area. I didn't get anything, but then I realized that I was seeing a picture, a scene. There was a clearing in the woods, with a fallen tree lying across it. I think that's where 131 and his family are."

"Did the wolf show you a picture of himself, or his mate?"

"No, but I did ask him what his name is. 131 is kind of a funny thing to call someone. I think he calls himself something like Charak, but I'm not positive. Now that I have a name it seems presumptuous to use it, since he didn't seem that thrilled about meeting me. Maybe I'd better stick to 131 until I know him better. Did you get any picture of him?"

She told me he'd shown her a brown and gray wolf, but she wasn't sure if it had been him, or 486, his mate. Whoever it was, he'd struck her as being on the small side, maybe just 75 or 80 pounds, and rather thin. I had been reading about Mexican wolves and told Raphaela that this is about the right size, Mexican wolves being quite a bit smaller than gray wolves. Mexican wolves, *Canis lupus baileyi,* are the smallest species of wolf and are about the size of an adult German shepherd dog.

"Is he getting enough to eat?" I wondered.

"I think yes, but he's not exactly feasting. He showed himself eating a little animal—I couldn't quite tell what it was, a little brown animal with big teeth, some kind of rodent. The wolf has one in mind he'd like to eat. He showed me the little guy sticking his head out of a hole, and then the wolf himself darting at the rodent but not getting him. He's hungry right now."

"Did you ask him how he feels about the reintroduction project?"

"I did, but he didn't seem like he knew anything about it. I tried to explain it to him by telling him about Wolf Haven, and asked him if he remembered it there. Finally he said that he did, so I asked him how he would compare his life there to his life now. He said something like, it's harder now but . . . and then he just conveyed by feeling that it is infinitely much better now. All that free food, the road-killed deer they used to give him at Wolf Haven? It's nothing to him, less than nothing, compared with freedom."

"Did you ask him anything else?"

"I asked what he thought of people. He indicated he really didn't have an opinion. I pressed him a bit and he said, 'I don't get people. I prefer to avoid them. Here where I live now, there aren't so many. That's good.'

"Then I told him that the reason I had called him was because I am an animal communicator, and I'm trying to show people that they can communicate freely with animals if both of us are willing. He seemed to find this mildly interesting, but it still didn't grab him very hard. His life now is what's on his mind.

"I couldn't help thinking of the wild orcas we talked to, especially Granny. She and 131 are both wild creatures, but it's quite a contrast. She is always so willing to engage with us, and he's so reluctant. But then, she has her life under control. All her family traditions are still in place, and above all she and her family have enough to eat. 131 is a pioneer. His mission is to recolonize the southwest for his kind, not to sit around shooting the breeze with the likes of me.

"I asked if there were anything at all he'd like me to pass on from him to other humans, and he said just that we do need to learn to share the world. That was about as much as I could get from him."

"Did you ask him anything about his first family? You know, 127, his first mate, the one who disappeared?" I asked Raphaela.

"I tried to, but it seemed like he didn't want to talk about it. I felt I was way overstepping my boundaries," Raphaela said. "I felt intrusive. When we call him this time, I'd like to start by doing a very formal greeting. Give him our respect and admiration, and tell him that we want to talk to him because he is a wolf hero."

"It's true, he is," I agreed. "Maybe with both of us tuning into him at once, we'll be able to engage his attention." We each put down the phone and quieted ourselves in preparation for making the contact.

As always happens when I am allowed to join Raphaela on a telepathic quest, the response was strong and clear, very different from the elusive contact I'd made the night before. This interview with 131 was also different in another way from any I'd experienced before. It seemed to take place not within my mind, as telepathic contact usually does, but out in the forest, in the wolf's home territory.

As soon as I'd called out to 131, Raphaela and I seemed to be standing together in our spirits in a clearing in the forest, the same one I had seen the night before. A compact gray and brown wolf stood in front of us, looking at us both curiously.

"So, you've both come?" he said. He seemed to be struggling to remember something. "Is it going to be like before?" he asked. "Like in the old days?"

I looked over at Raphaela and found I could see her quite well. Her spirit was large, blue, and impressive. I could see my own spirit too and it was smaller and a lighter color, more green than blue, and generally less substantial. Then I looked inquiringly at the wolf.

"Yes, perhaps it will be like before," he decided. "People used to come . . . and now you come again. I'll talk to her." He indicated Raphaela. "You can be a listener. It's fine to have an observer, but I will tell her what she wants to know."

The two of them began carrying on a conversation, and I was able to catch some drift of feelings from his mind as they talked. I gathered from this that what he and his family are mainly doing is looking for a pattern to fit their lives to. Wolves, he felt, used to be part of a pattern in the land—I understood him to be thinking of a kind of design in nature, a series of interrelationships with other animals, with the seasons, and with the land itself.

He thought the places in the pattern where wolves were meant to go were still there, but he wasn't sure. For him, the pattern would be found in the cycle of his year—where he and his family would go from season to season, what they would kill and eat, where they would go to den. He also called it their "routine" of hunting and eating. I understood that he loved to have "big people" (elk, deer) to eat, and he also appreciated the "small people" (I think he meant the little brown rodents he'd showed Raphaela the night before). I remembered the wolves in Farley Mowat's *Never Cry Wolf.* Despite their reputation for bringing down caribou, in lean times they subsisted mainly on mice.

Several minutes went by. I waited quietly until 131 finished talking with Raphaela. Then felt I had his attention again, as if he had suddenly remembered something he wanted to tell me. He looked straight at me, and said something like, "Wait—this is part of the pattern too. We used to meet human people in spirit like this in the old times. People like you, and people like me, would meet sometimes and talk. Do you remember?"

I looked back at him, something stirring in my memory too. "You used to give us something," he went on. "Something pungent. What was it? Blood? Pee? Smoke?" Both of us hung our heads. We felt ashamed of not being able to recall something so basic and important.

"Well," he said. "Let us stand here before Spirit. She will give it back to us, but we must stand here and listen. We'll wait."

About this time Raphaela came back on the phone from her conversation with the wolf. I told her what I'd learned so far, and asked what she and the wolf had been discussing.

She said he had still seemed cautious and guarded. He wasn't hostile, just remote. He said something like, "You again?" She tried once more to explain to him what we were doing by saying that she was writing a book that would help people know and respect animals, including wolves.

He said he didn't "get" books—he didn't understand what they were. Raphaela explained it as a way of spreading information, something like howling. That perked him up. "Howling!" he said. That was thrilling and important. Very important, very big.

Then Raphaela remembered that she had planned to greet him more formally when she made contact this time. She had forgotten to do it at the beginning of their talk, but she did it now. She bowed before him and acknowledged that she and I were visitors and that the land was his. It seemed to be well received. The wolf gave back a feeling that maybe she was all right, since she knew how to show respect.

We had now reached the point when the wolf told me that he would prefer to speak with Raphaela, and I could listen. To her, he said that while he did not much like verbal communication, he would be willing to show her pictures of his life. He then showed her a series of pictures around the clearing with the fallen tree across it, which I had been shown the night before. This was where he and his family were living. The land was rocky and steep, touched in places by patches of snow. The sky was the intense, deep blue of higher altitudes. The air was hard, crisp, and very cold. He was aware of other wolves in the area, traveling, and of the people in the sky—eagles, hawks. His visual acuity seemed to be excellent, for the pictures of the flying raptors were sharply detailed. Raphaela got the impression that the birds were very important to him and might bring him information about the area.

Raphaela thanked him for the pictures, then asked about his family. "Yes!" he acknowledged. His family was with him. The pups that had been born at Sevilleta were half-grown now—somewhere around 40 pounds. His attitude towards them was positive and very elemental: something like, *the kids are good. They will learn to help us.*

Now, he said, the kids mostly played. They frisked about. He showed them frisking about. He did not frisk. He was serious and businesslike. The kids would be that way soon enough.

He expressed the same sort of feeling for his mate: *She is good. We help each other.* Raphaela said it reminded her of a farm family: the children were loved, but they were also respected and needed as hands around the place. This wolf family was very much in business together. They had a living to make.

She asked them what they were eating. He showed her bird feathers! Did I think they could they have eaten a bird? I had no idea. He also showed her a picture of a wild pig. I knew from reading field notes on the Fish and Wildlife Web site that there were javelina, wild pigs, in the wolf restoration area. The wolf said these people were hard to catch, vcry hard, but they had done so! Not often, he admitted, but they had. He was quite proud of himself.

Next Raphaela wondered if they had been getting any supplementary food from their human sponsors in the Department of Fish and Wildlife, who had supplied them with road-killed elk and deer while they were in the acclimation pens. She didn't get anything from him suggesting that they were. They seemed to be on their own.

What do you do for water? she went on. Right now, he said, we're eating snow. And in the summer? He showed her some muddy seepages. It didn't look great but the hardship didn't seem to interest him particularly—it was just how things were, not that big a deal.

Was there anything else he'd like to tell her, she asked, any significant events or features of his life she hadn't thought to inquire about? What was the real essence of his life? He responded that hearing other wolves in the area was very, very important to him. So was traveling. That was freedom, to him—the ability to go, not just around and around in a cage but to *go,* to *move,* to *cover ground.*

As for the essence of his life, it was *being himself.* Being a wolf. Now, for the first time, he spoke in words. "It's hard. It's simple." He turned to look at her, and his eyes blazed out with wolf fire. Raphaela caught her breath. "It was like the time you saw the Great Cat," she told me. "The wolf opened himself up and showed me what he really is."

When Raphaela related this, I too caught a sight of his inner nature. The light blazed out of his eyes. I've heard the light that comes from a wolf's eyes referred to as "green light." To me it was a yellow fire, streaming from his eyes. He had a huge, overwhelming beauty.

The wolf faded away from us; the interview was over. That night I lay in bed and thought about my wolf. I wanted to call him again, and then I thought, "My wolf?" What was I thinking? He wasn't my wolf. He belonged to himself. I could have no possible claim on him, and he had very little interest in me, but I longed to know him better. He was a being like no other I had experienced—wilder and freer.

He was wilder than the wild orcas I had encountered, and also more alien. I wondered why that should be. He was, after all, a close relative of Theo and Julie, my dogs. Also, he was a land mammal, not a sea creature like the orcas. Yet he seemed far stranger and more remote than they did.

I think perhaps the difference is his role as a wolf pioneer, and the desperate, highly focused quality this gives to his life. Granny and the other orcas lived in a stable culture in the familiar waters of Puget Sound. They were ensconced on top of the food chain; they had enough to eat; they had time to talk and think and socialize with us. They had relatives all around them—their own close family in J pod, their cousins in K and L pods. All that makes them emotionally accessible to us. They feel like kindred spirits, like people we might encounter anywhere, who just happen to live in the water.

131 was different. He was essentially alone out in the wilderness, and of course he was alone because my kind had been so brutal and unfeeling to his kind. That was the harsh truth that underlay any encounter between us. It was not surprising he was elusive and wary; it was amazing that he consented to talk to Raphaela and me at all.

However, in the perverse way in which one always wants what is rare and elusive, I wanted to be with him, to know him. Each night for the next several nights I called out to him (receiving only brief replies, or nothing at all) and thought about him all day.

Allusions to wolves began occurring—a sure sign you've become sensitized to a topic. I read an article that mentioned, in an offhand way, that Mount Gabriel, in West Cork, is the place where the last wolf in Ireland was shot. A friend told me about an encounter he had had with some half-wild wolves on a trip to Brazil a few years ago. He was staying at an eco-tourist camp in the rain forest, and walked out onto the grounds with a Brazilian guide who said, "I'm going to call the wolves now." The guide made some wolfish noises, and two or three long, thin wolves had come out of the forest. The guide gave them some food, and they slipped away again.

I got the idea of asking my dog Julie for help. As a relative, however distant, of 131, she might be able to advise me about the best way to contact him. I had read that wolves in the wild do not want dogs around; they consider them interlopers in their territory. There was a Department of Fish and Wildlife directive which firmly stated that you were *not* permitted to kill a wolf for attacking your dog (you should leave your dog at home if you planned to camp in the wolf restoration area.) But it might be different in the spirit world, and I thought it couldn't hurt to ask.

Julie was indeed helpful. She reminded me of what 131 had said—that people had once come to see him, and that we had given him something. She thought I should pursue that, and try to remember what the offering had been. She also said that while she, as a dog, lived and worked close to me and helped me keep the integrity of my home, wolves actually used to help people in a different way. They were not close companions to people, but they did give them spiritual help. She thought I should let go of my focus on the recent, troubled relations between humans and wolves, and try to get back to an older, more stable relationship.

I was grateful that Julie, calling on a longer memory than mine, had taken me back to a time when humans lived in harmony with all animals, not just with our domestic animal companions. I longed to have 131 in my life on any basis he would allow, and resolved to keep looking for the right way. Then I remembered Jasmine and her whale, and that gave me an idea.

Jasmine Indra is one of the animal communicators who went with Raphaela on the humpback whale trip, aboard the *Bottom Time.* She's also the human companion of Benoji, the black cat who helps her communicate telepathically with all beings (and who memorably advised her that *she*, at least, would give little Jazz a smack). I recalled her telling me later that a whale had become her special friend and guide. I called her to find out what had actually happened.

It seems that Jasmine's encounter with her whale, Jenny, began on the trip in the *Bottom Time.* Teresa Wagner, the trip's official animal communicator, offered each of the people on the trip a "whale transmission"—a session in which Teresa would ask the whales to provide that person with their own special whale guide. The whales seemed happy to be part of this project.

When Jasmine had her turn, Teresa told her that the whale who had volunteered to be her guide was called Jenny, and she was a baby humpback who was actually present with them at Silver Banks.

Jasmine asked whether she had seen Jenny, and Jenny said that Jasmine hadn't seen her yet, but she soon would. When she did, Jasmine would know that it was she. She also said that although Jasmine didn't know her, she already knew Jasmine and loved her. Her first words to Teresa were, "Oh, Jasmine! Oh, Jasmine!" spoken in a loving, open way, like a little girl.

The next day, as usual, Jasmine got into one of the small boats to go out and look for whales. They had just pushed away from the *Bottom Time* when a group of dolphins came up to the boat and began playing the bow wave. They seemed to be leading the boat. Jasmine soon saw that they were leading them towards two whales a little ways off. As they approached the whales the dolphins left the boat and swam over to the whales.

The two whales were a mother and her baby. When the dolphins reached them, the baby left her mother and swam towards the boat, then dove. Suddenly she spyhopped, putting her head out of the water not four feet away from the side of the boat. Jasmine felt that this had to be Jenny. She got into the water with the whale, and the two of them looked at each other, just feet apart. She stayed just a little while, and then swam back to her mother, and the two of them swam away.

"Thinking back on it, I'm sure it was Jenny," Jasmine said. "I'd been mentally calling her the whole time, I was so anxious to see her. At the time I did have a few doubts, but I'm always struggling with self-doubts—that's one of my issues. It was definitely her. For one thing, no one else on the trip remembers the encounter particularly. I'm the only one who noticed it as something really, really special, which I think is because it was meant for me. We had many whale encounters on that trip that were more meaningful to other people, or which lasted for much longer, but I never had one that felt so intimately connected with me. And it was the only time on the whole trip that I saw dolphins."

Later on, when she was back home, Jasmine called Teresa and asked her if she could tell her anything else about Jenny. Teresa told her that Jenny said she would always be there for her, and would help her in any way Jasmine wanted.

Now, Jasmine feels that Jenny is always there. It's as if she is actually part of her own consciousness. She added that before going on the trip, she had found a card that she liked a lot. It showed a little girl with another little girl, her friend, standing behind her. The girl behind had angel wings on her back. She brought the card with her on the trip as an inspiration. She feels that Jenny is like that to her—a special friend, a guide, and an angel.

Jasmine's story inspired me, in turn. I saw it as a direction that I might take with the wolf. As a human, a person with a body, I could have no place in his life. Human presence has almost destroyed wolves; the only thing we can do for them now is to get out of the way and allow the environment to heal.

But our spirits aren't bound like our bodies are—this is the clearest message of animal communication. A wild humpback whale in the Caribbean can become the friend of a woman in

Berkeley. Perhaps a wild wolf in the mountains of Arizona could accept some friendship from a woman in Eureka. Now I just had to find a way to approach him.

The Spirit Wolf

"Are you thinking about our wolf as much as I am?" I asked Raphaela a few days later. She nodded agreement—he'd affected her strongly, too. "Should we try to contact him again?"

"We could, but I have another idea," she responded. "Let's call his first mate, the one who disappeared. No one knows what actually happened to her. I'd like to know, and 131 doesn't seem to really want to talk about her."

I readily agreed and we prepared ourselves to make the contact. It felt strange calling a wolf and not knowing with certainty whether she was alive or dead. I was pretty sure she had to be dead, since she hadn't been seen since September 1998; but there was that little bit of doubt, or hope.

127 came in when I called, but even then I wasn't sure. I saw her as a giant wolf image floating over the whole area—like a range of clouds over a range of mountains. Afterwards Raphaela explained that animals often appear liked that after they have died. In fact, if an animal dies in your presence or while you are telepathically connected, you can often see the animal expand until he or she fills the sky in just the way I saw. When I saw her like that, I should have known she was dead.

I didn't, though, so I was still in doubt. However, I knew enough by now to know it didn't make that much difference. I approached her—she was very beautiful—with respect and just said that I was there to visit her if she would allow it.

She responded with a strong flash of anger. She said, "Why should I even talk to you?" I replied that for all that had happened to her and her family, I was very sorry. As quickly as her anger had flared up, it died down, and she asked me in a kind way what I wanted.

I said that I wanted to know if she remembered anything about the ceremony with which my ancestors used to greet hers. I said that if I knew the right actions I would perform them for her and her family. She said with great certainty that the ceremony had been putting fat onto the fire. But, she amended, that was only the outer form of the ceremony. It is really an inner event. The fat is just a symbol. The real substance is that I—a human—agree not to take all the fat of the land. By putting it on the fire I am saying, I will share. I will leave some. I will not take it all. I will take just what I need, but not more.

Thoughts of all the things I have taken from the land that I don't really need streamed through my consciousness, and I felt profoundly ashamed of myself. The wolf seemed aware of my thoughts. She said, "Yes, this is reality for you. This is what you can do for me. Take less. Eat the small things mostly."

Then she changed the subject and said, "Some people will choose us. You may not. There are so many animals, and some people will feel close to us and some will feel close to other kinds. We know much more now than we did. In the old days, when the ceremonies were in use, we only knew a few kinds of animals, and so did the people. Now we know there are whales! We didn't know that before. It is the same time for us as it is for you—a time of travel when knowledge expands and takes in the whole planet."

I wondered how she knew about whales now, and then remembered that she had been born at Wolf Haven, in Tenino, Washington—not far from Puget Sound, home of our dear orcas. Wolves are great natural travelers, but even so, this was unusual!

Just then I heard Raphaela returning to the line from her conversation. I said goodbye to the wolf and asked Raphaela what she had learned. "Is she alive?" I asked first.

"Oh, no, she's dead all right. But she said it was a glorious, glorious year, her year in freedom. She was rejoicing in her freedom the entire time. Then, she said, she was shot just like the others—that would be her children. The only reason her body hasn't been found is that her pelt was taken as a trophy. It's hanging on someone's cabin wall right now."

This was sad but not really surprising. I told Raphaela about the flash of anger I'd felt in her.

"I felt she had a certain bitterness," Raphaela said. "When she told me about being shot, she said, 'It's just the fate of wolves.' But she also had some hope. It helps her a lot seeing the wolves who are still out there. She watches over them. She's doing whatever she can to help them make it. I wonder—when 131 told you that you would ask the Spirit about the ceremony, did he mean her? She is a spirit wolf, after all."

"Did she tell you anything about the ceremony?"

"Yes, and she did remember it. She said it was putting meat on the fire."

"Really! I got almost the same thing. I thought it was fat, not meat, but it was supposed to be put on the fire. What else did she say?"

"She said that the people would stand in a ring around the fire. The wolves would stay out of range, but they participated, they were part of the ceremony. She said, 'Someone said favorable things about wolves, and we accepted the offering. They ate, and then we ate.'"

Raphaela had also been told that the offering was symbolic. "We couldn't live on the little bit of meat they gave us," the wolf pointed out. "But it showed that they were willing to share. That was the meaning."

We were both filled with reverence and gratitude to this lovely being. I realized that she had given me a picture I needed: a vision of wolves and people sharing the land. It was possible; it had been once and it could be again.

Chapter Eight

Animal Communication

Raphaela: How to Communicate with Animals

I hope I can inspire you to begin communicating with animals yourself. Teaching this skill is a big part of my life and I love doing it almost as much as I love talking to animals. But really, you know, it isn't actually *teaching.* It's more like helping people *remember* a skill they had as a child, but later lost.

You too may find that you used to communicate with animals in the past. Sometimes the memory is so deeply buried that it takes a while to get back to it. Often there is a hurt somewhere in the background—some sad event that turned you away from communicating with animal friends. In Sonya Fitzpatrick's book *What the Animals Tell Me,* she tells how she put away her ability to talk to animals after the death of her beloved geese friends. If something like that happened to you, you are going to find that beginning to communicate with animals again will help heal whatever it was that stopped you before.

Whether you remember talking to animals as a child or not, what's important is just to jump on the bandwagon and start doing it now. Don't dwell on the past—animals never do! They just enjoy the present.

It is amazing how much more you will enjoy your animals if you do start communicating with them. I have certainly seen this over and over. You think you have a close, loving relationship with your animal companion, and of course, you do, but you will be fascinated to find it becoming even closer and more loving when you open up this channel.

I saw this happen while working on Chapter 2 with Elizabeth. She told me that after she began to understand more about her Shi' Tzu, Theo, and his background as a Tibetan monk, their relationship blossomed. He'd always been a sweet, loving dog, but there was also a certain reserve about him. Now it seemed as if being more fully known or understood had liberated another side of his personality. He became livelier, more active, and even more adorable than before. There was a deepened quality to their interaction that she treasured.

A Quiet Mind

Now, let's get you started. As I always tell people in my workshop, animal communication requires a quiet mind. This is easier said than done. Our minds are naturally active. We go from thought to thought, scarcely ever experiencing the quiet space where the thoughts actually come from.

However, mind chatter is one of the two biggest obstacles to full two-way communication with animals. (The other one is self-doubt, which we'll come to in a moment.) Regular meditation, yoga, qi gong, or any similar mind and body practice that increases sensitivity while quieting the mind will help. Spending quiet time in nature is also good. Elizabeth has asked me to be sure to mention Transcendental Meditation, which she has practiced for many years. TM is simple, natural, and easy to learn, and regular practice will certainly help you with this very necessary step.

Here are a couple of very simple meditations I use in my workshops. If you spend 15 or 20 minutes doing one of these each day, you'll notice an immediate reduction in your mind chatter.

Sit quietly and comfortably where you won't be disturbed. Take a few deep breaths, and concentrate on releasing tension and effort as you breathe out. You might want to roll you neck and shoulder muscles a few times too, further releasing tension. Tell yourself, "Nothing to do, nowhere to go." (At least not for the next 15 minutes!) Then begin focusing on your breath. With each in-breath you can silently say "In," and with each out breath silently say "Out." If your mind wanders, simply say to yourself, "Chatter, chatter," and return your attention to the breath.

That's it. That's your entire responsibility for the next 15 minutes. Nothing difficult or weird or hard to master. Even this simple meditation will lower your blood pressure and your heart rate, and slow your thoughts down to a manageable level, making a quiet space for you to hear the animals.

A similar simple and lovely meditation is to again concentrate on the breath. This time keep your attention entirely on the tip of the nose, where the breath comes in and goes out. Don't follow the breath into the lungs. Just keep your attention on the tip of the nose as your warm breath goes in and out. Again, if you find your attention wandering, simply return to the breath. Don't get mad or be upset if your mind wanders. It will wander. It will shop, cook, clean, pick the kids up from soccer, and make bread, all in less than a minute. That's the nature of the mind. Your job for the next 15 minutes or so is to simply keep returning your attention to your breath at the tip of the nose, each and every time, without exasperation or even comment.

Sometimes it's helpful to count the breaths up to ten, and then start over again. It's a little trick to keep the mind busy while you meditate. Sometimes the rhythm of the breath may remind you of waves washing in and out on the beach, or a gentle spring breeze sighing through an orchard.

Gradually, over weeks or months, you may find meditation becoming the most pleasurable and restful part of your day. You may look forward to it, and crave the spaciousness and quiet it brings into your life, a quietness essential to sensing the thoughts and feelings of your animal friends.

Say Hello

Now, I'm going to assume that your mind is clear, quiet, and restful. You aren't unduly preoccupied or stressed, and you have put any preconceptions or judgments on hold. The next step in animal communication is really easy and fun. You are going to throw out a "Hello."

One of my students told me she visualized her "Hello" as a cartoon balloon above her head, and now I see it that way too. So imagine that your "Hello balloon" is out in front of your chest. Now throw it up into a corner of the room, and see how that feels. Next, imagine throwing it out the doorway of the room or building you are in, and up to the top of a tree or telephone pole. Notice how that feels.

Continue practicing by sending out a greeting to the roses in your garden, or the shrubs in a neighborhood park. Then try something further away, like the redwood trees in Northern California. How about the trees in a Brazilian rain forest? Say hello to them too.

While you're experimenting with this, notice any impressions you get. You may find that you naturally and easily start to bring in information from the plants you are tuning in to. Yes—the plants are responding! They are sending you back their own hello.

The next step is to choose an animal friend, and send your hello out to this animal. The animal can be in the same room with you, or hundreds of miles away. Experiment with sending your hello both to animals close to you, and to ones who are farther away. They will feel different, but rest assured, your greeting arrives to both. As you practice you'll begin to see that it doesn't really matter how far away an animal is, even though it may feel a little different.

Again, as you send out your greeting, keep your perceptions open for any impressions you may receive back. Don't worry if you don't get anything at this point. You may or may not, but you're getting yourself ready to move on to the next step.

Now for the tricky part: receiving. It's all very well to talk *to* animals. Lots of people do that. The fun part, the part that makes this a two-way street, is when *you hear the animal respond.* What I want you to do is *to imagine you feel, sense, see, or hear a hello coming back.*

That's right, imagine it.

Use your imagination to let the process start and become real for you. If you cannot imagine and conceive of telepathic communication as a reality, you will miss the experience completely. So go ahead and imagine that you hear, feel, or sense a "hello" coming back from your animal friend. It may be crisp and clear or you may have a softer impression of the animal. Whatever it is, just accept and enjoy it.

"Using your imagination" is rather different from what most people, well, imagine it to be. Imaging something doesn't mean "just making it up." Rather, you are using one of your mind's most creative faculties, whose function is to turn subtle impressions in the depth of your mind into concrete, fully conscious *images. The Oxford American Dictionary* says imagination is "the ability to imagine creatively or to use this ability in a practical way," which I think is quite good.

In my workshop I often start with a game to strengthen the imagination. I wrap up a little present for each participant in such a way that they all look exactly the same—there's nothing in the shape of the wrapping to give away what might be inside. I put all the presents in the middle of the circle and each person takes one. Then we go around the circle, and each person is asked to sit quietly and imagine what might be inside.

You'd be amazed how often the imagination is exactly right. I remember my very first workshop, in Elizabeth's barn. I asked Elizabeth, as our host, to go first. She selected a present—like all of them, a wrapped cylinder about eight inches long and two or three inches around. She held it up and stared at it. "Some kind of knot or bow," she guessed. She opened it. It was one of those chew toys for dogs, made of strands of colored rope tied into a knot.

Next, her friend Julie Clare held up hers. "Catnip," she speculated. Inside she found a catnip mouse. The next guessed, "Something you play with. White." It was a white ball. And so it went, around the class. Even I was surprised when everyone in the class guessed either exactly, or quite close, what was in their present. It was my first class, and I didn't know then that this would be the rule, not the exception.

The most interesting one, however, happened after the class was over. I had a present left over and I offered it to Elizabeth's husband Ralph, provided he would guess what was in it first. Now, I should tell you that Ralph is a kind, tolerant man who loves Elizabeth and respects her views. Having said that, he is definitely a nonbeliever. He's an engineer by profession and a scientist by training and temperament. When it comes to believing in telepathic animal communication, he simply, respectfully, declines.

When I asked him to guess what was in his present, he said, "How would I possibly know? A brush." Afterwards he explained that he said that because a brush was the most useless thing he could think of—he is perfectly bald. Yet, when he opened it up—a brush is what it was.

Did this experience make a convert out of him? It did not! But it gave me a wonderful illustration of how the imagination works. When Ralph took his guess, he was using his imagination in a light, humorous way. It doesn't matter that he thought what he was saying was slightly ridiculous. Maybe it was, but it was also exactly right! And that's the point. The same thing will happen to you. Just toss out your hello to an animal, and then imagine that the animal has said hello back. Use your imagination—and you will be right.

Starting the Dialogue

Now that you have opened yourself up to what your animal friends have to say, you're ready to start a two-way dialogue with them. I suggest you begin by asking simple questions.

Remember that your mind does need to be quiet for this to work. The answers you will get to your questions will come to you quietly. If your mind is racing around from one thing to another, you may not hear the reply at all.

Many of my students worry that what they are hearing is not coming from the animal, but from their own mind. That's exactly why you have to be able to quiet your mind. If your mind is clear, still and receptive, and you have no agenda, the thoughts and information that you receive really will be from the animal.

I suggest that before you begin your first session of two-way dialogue, take a few moments to go within and contact the quiet level of your mind you have gotten to know through your practice of meditation. Then send out your first question to an animal.

At my workshop, people generally bring their animal companions with them, so there are plenty of animals to practice on. I have students start by asking someone else's animal companion a question like, "What is your favorite food?" This is a good question because the animal's human companion is right there to confirm what the animal tells you. Other good questions are, "Who are your animal buddies at home?" "What do you like best about your environment?" and "What is your favorite activity?"

The answer may come as pictures, words, thoughts and feelings, physical sensations, or sometimes a whole concept or gestalt. Your job is just to accept whatever comes, whether you think it's "right" or not. The answer usually comes quickly, too, so don't belabor it or stand around waiting for lightning to strike. If you hear a word or two or have a brief feeling, *that was it.*

I remember once, in a workshop in Denmark, one of the students was talking with a llama. She asked the llama his favorite food, and got the answer "Bananas." She thought that was so unlikely and outlandish an answer that she almost didn't share it with the group. But she plucked up her courage, and reported what she thought the llama had told her.

"Oh, yes," the llama's person responded. "She adores bananas. They're her favorite treat." And this was true not only of this llama, but of most of them! Turns out that what apples are to horses, bananas are to llamas. Who knew?

Overcoming Doubt

We have now arrived at the second great obstacle to animal communication: self-doubt. Once you have become sufficiently quiet inside through meditation to hear animals speaking, self-doubt is literally the *only* thing that stands in the way of your enjoying full communication with all beings.

Doubt is natural enough. Everyone feels it. I certainly did at the beginning. It's hard not to, when so much of the environment denies the reality of telepathic communication. But you must get past your doubt—it's simply a waste of time. The best way is simply to practice. Whenever possible, get concrete confirmation that what you have received telepathically is correct. This will remove your doubt quickly.

When the llama's person confirmed that my student in Denmark had heard correctly, it was a very powerful and helpful experience for everyone at the workshop. At another workshop in San Francisco, a student who thought she wasn't receiving anything at all from any animal asked my Golden Retriever, Tootsie, about Tootsie's animal friends at home. The student's brow wrinkled as she received a picture of "a really skinny small brown and black dog." I laughed as I recognized her description. Tootsie had sent her a picture of F'lar, our extremely lean brindle whippet. Skinny is right—Tootsie had described F'lar perfectly. The result was that the student's confidence blossomed.

Sometimes animals who are not physically with you may send a picture of themselves when you first tune in. If they don't do so spontaneously you can ask them to. Then you can use that picture to verify that the communication you receive is really coming from that particular animal. Once I was speaking to a horse for the first time, and she showed me that she had a refined, aristocratic face with a beautiful long narrow blaze. I asked the horse's person, "Would you describe your horse as refined-looking, or more on the stocky, robust side?" The person's answer told me I was indeed speaking with the right horse. Sometimes you may call one horse and another horse nearby may come in—it has been known to happen. Checking for confirmation is a good idea.

During another consultation, a client asked me to speak to her dog's litter of eight identical white Bichon Frise puppies. The person already knew the personality of each and every puppy, and she wanted me to check to be sure I was speaking to the right one. This made for a bit of a headache for me. The person would say,

"Now it's Mopsy's turn." I'd tune in and get an adorable little white ball of fur who seemed quite lively. "I think I have Mopsy. Is she a lively little puppy?" "Oh, no. You must have Popsy. Mopsy is much more mellow. Try again." I think I eventually got the whole litter, but my head was reeling at the end. The physical-appearance method of confirmation does have its limits!

Another very reliable method you can use to confirm that you are tuned in is to ask the animal for physical verification that the animal has received your communication. You can only do this when you are actually present with the animal, of course. Remember I told you in the introduction how my dog Petey walked across the room and put his paw in my lap when I asked him to demonstrate that he was really hearing me? That is what I mean by physical verification. It's very helpful when an animal does this. It provides objective "evidence" that you are really tuned in and that the animal is listening.

However, you may not always be able to get your animal to play along. Animals do vary quite a bit in their willingness to provide physical confirmation. While Petey would practically tap out Morse code for me (he was just an angel that way and very eager for me to get it) I have other animal friends who regard physical confirmation as a pathetic concession to human insecurity. If you ask for it all the time the animals may start to feel controlled and manipulated, and will resent it. They want you to participate in discussions as an equal, not as an underling needing constant reassurance! I recommend that you ask for physical confirmation just once or twice at the beginning. Then your own confidence should take over.

A year or two after Petey died, a wonderful cat, Sophia, came into my life. I had decided to get a kitten from the Berkeley Humane Society. I went and looked at all the kittens, but hadn't decided, when the attendant came up to me and said, "You don't want any of those kittens."

"I don't?" I said, surprised.

"No, you want this one." She led me out of the kitten room to her desk in the reception area. There, in an open drawer, sat Sophia. She was about nine weeks old, a gorgeous light fawn with faint brown tabby stripes and icy blue eyes. She had little tortoise shell ears and brown velvet paws. She reminded me of Ling, my beloved childhood cat, and I was instantly smitten.

I tuned in to this adorable kitten and asked her if she wanted to come home with me.

"Of course," she said. To confirm that I was really tuned in to her, and not picking up messages from the other kittens, I asked her to look at me. She gave me one pitying glance, then looked away and said, "You're not in kindergarten anymore."

Sophia has steadfastly refused to give any physical confirmation whatsoever since that initial withering glance ten years ago. She feels it's beneath her, and apparently she feels it's beneath me, too!

Jasmine Indra, Benoji's person, had quite a time with her cat family when she began seriously studying animal communication. Not a single one of her four cats would give her the slightest physical sign that they heard her attempts to communicate with them. Finally, after weeks of sweet supplication, Jasmine switched to tough love.

"Okay," she said to the cats one evening, "No dinner until we've had a little chat, my friends." Jasmine explained to the cats (for perhaps the hundredth time) that she was studying and practicing telepathic animal communication and needed her cats' help and acknowledgment. She would give them dinner only when they gave her some sign they could hear her communication.

It wasn't until quite late that night that each cat finally gave in and responded to her physically. One gave a deliberate glance, another a long slow blink, both signs Jasmine suggested to them. Then the meal schedule returned to normal.

Sometimes people don't think twice about the concept of telepathic communication once they've been introduced to it. They just plunge in, expecting it to work. It usually does. Years ago I did a consultation for a woman named Joan Rogers. Her beautiful German Shepherd dog, Helga, had a severely lacerated ear that required stitches and antibiotics. How, Joan wanted to know, had Helga gotten injured?

I tuned in and asked Helga what had happened. She gave me a description, partly in words and partly in pictures, of digging under the tall chain link fence surrounding her property. She had dug down maybe 12 or 15 inches when she encountered a wad of old rusty barbed wire twisted together. With her head deep in the hole, and her passionate digging, she somehow managed to get hooked by the wire. Struggling to get out of the hole, she tore her ear.

I explained this to Joan. "Right," she said. "Thanks for the information."

Then (as she told me later) she hung up the phone and turned to Helga. "This is ridiculous," she said. "If Raphaela can talk to my dog, *I* can talk to my dog. Okay, Helga. Where is this barbed wire?"

Helga instantly turned and trotted out the front door, through the garden and down the path to the fence. There Joan found the hole and the barbed wire embedded in the dirt at the bottom.

Joan kept me busy for months after that, talking to her race horses, riding horses, brood mares, and stallions, not to mention the many Burmese cats and German Shepherd dogs in the family. Eventually she took my workshop and now talks to all her animals herself.

How It Will Come

Animals send information telepathically in many different ways. Sometimes you get the animal's feelings and emotions about a situation. Sometimes animals show you pictures, or give a combination of words, pictures, feelings, and even physical sensations. Occasionally animals send you a whole gestalt, something like the "ball" of information that Jasmine's cat Binoji described. Here are a few examples of how the information may come.

Once I was doing a pro bono consultation for a woman who rescued Doberman Pinscher dogs. She had asked me to tune in to an older Doberman she had rescued months earlier from the pound on the day he was due to be euthanized. She couldn't keep the dog at her home, and had no alternative foster home lined up. So she had rented space in an out-of-town kennel where the dog received food, water, and shelter, but absolutely no individual attention.

When I tuned in I was overwhelmed by this dog's feelings of darkness and despair. I didn't perceive anything else for several moments. Finally, out of the dark, I heard the words, "I'd rather die than spend another day in this place."

I was shocked. I had never talked to an animal who was so depressed and despairing. Usually, no matter what the circumstances, most animals seem accept their situation and even have a bit of hope or optimism. This dog did not.

I explained to the dog that he would have been euthanized if he had not been rescued. "Yes, I know," he said. "It would have been better than this."

We continued talking. The dog gave me pictures and feelings about his early life. He showed me that his person was a young man who took him to work with him in a pickup truck. All day he was with the man, who worked on construction sites. He napped in the truck or trotted around the construction site. He sent me feelings of pleasure and contentment as he showed these pictures. Then, abruptly, life changed. The dog was taken to the house of an older woman who I thought was probably the young man's mother. The man left without a word, and the dog never saw him again. He was completely bewildered.

Now the dog showed me pictures of being in a small back yard, filled with dog poop and flies. He sent feelings of loneliness and misery. He was never allowed in the house. No one walked him or talked to him. Sometimes two days would go by without food or water. Finally the woman took him to the pound. He never understood what happened to his owner. This and his present situation were causing his deep depression.

I passed this on to my client, who was horrified. She had certainly not realized how bad the dog felt. Initially, when she rescued him, she had intended to go to the kennel daily to walk and play with him, but had not been able to because of the distance and her own poor health. Now she realized how serious his condition was and promised to do better.

This kind of experience in animal communication, where an animal communicates feelings of great sadness, is fortunately quite rare. But it can happen, and it always makes me realize that communication by itself can't solve all problems. The dog needed daily care to lift him out of his depression; more than that, he needed a home. However, communicating with him did at least alert my client that he desperately needed some attention. I, for my part, prayed that she would be able find a wonderful home for him soon.

At the opposite end of the emotional spectrum, you may receive feelings of pride, pleasure and elation. I felt all of these when I tuned into Windstrike, a beautiful chestnut Thoroughbred from Joan Roger's stable of race horses. I was dazzled by how confident and "up" he felt, and asked him why he was so happy.

Windstrike flashed me pictures of a race. As the pictures unfolded, he grew more and more excited. He showed me his huge stride and surging speed, and the turf flying in his face as he drove for the finish line, his lungs bursting for air in an all-out effort. Then I got his intense pride as he crossed the finish line first to win the race. I couldn't help experiencing his exhilaration and pleasure right along with him.

I came back on the phone with Joan, and reported what Windstrike had showed and told me. She burst into laughter. "He actually came in second on a photo finish," she said. "But we're not telling him!"

You can also get feelings from animals that are quite surprising. I remember a horse called Winston whom I was called to consult with after he did much less well than expected in two races. He was a three-year-old Thoroughbred, and his trainers were very excited about him. When they would let him run full out in training, and clocked his speed (an exercise called "breezing"), Winston's times were astounding. The trainers smelled a winner. But when Winston was entered for his first race, he hung back behind the leaders, and finished with a much slower time than he'd clocked when breezing.

In his second race, the same thing happened, and his trainers watched him closely. They noticed that he was definitely holding back. There was no question in their minds that he could have passed the lead horses—for some reason, he just *didn't.*

I asked Winston what was up, and he said, "Yes, I'm fast. But those mares!" And he sent me a feeling of mingled respect and fear. "Do you know what would happen if I passed them? They'd beat me up for sure!"

You may also experience an animal's physical sensations, particularly if you are asking the animal questions about where he hurts or how he is feeling. It can be a little unsettling if you are not prepared for it. If this happens, it's important that you release the feeling at the end of the conversation, and not take it on yourself.

When Windstrike described the race to me I did actually feel his huge stride, the chunks of turf flying off the track stinging his eyes, his speed, and the burning of his lungs at the finish line. I felt my heart rate increase, and adrenaline surged through my body. I felt giddy with exhilaration, and had to sit quietly for a while to get back to normal after the consultation.

Releasing the sensations is even more important when you communicate with a sick animal. When my sweet Neatherland dwarf bunny Daisy became sick I asked her what was wrong and how she was feeling. A sensation that I was fighting for air came over me. I felt fright and panic. I felt my head tilting up and found myself experiencing Daisy's panting.

When I realized what Daisy was going through, I rushed her to the Veterinary Medical Teaching Hospital here at UC Davis. The student vet doing the admission examination took one look at her blue mucus membranes and panting respiration and hurried her into an oxygen tent. When I checked in with her a few hours later she had relaxed and recovered her equilibrium a bit. Although she was no longer panicked and fighting for every air, she told me she still couldn't take a regular, normal-sized breath.

The next day the vet called to say that Daisy's chest x-rays demonstrated a pattern consistent with widely disseminated cancer. There were very few treatment options. Antibiotics and intravenous hydration could possibly prolong her life, but she would not be able to recover.

Together, Daisy and I decided that the time had come for euthanasia. I went to the hospital to be with her as she left her body. She sat on the metal examination table and licked the tears off my face as I held her in my arms one last time. A few minutes later the vet arrived and administered the medication releasing her spirit.

She was a sweet, gentle, beautiful spirit and I still miss her. It is so hard to let go of our beloved animal buddies, and hard know what is the "right" decision when considering euthanasia. Telepathic communication is really a blessing. The beloved animal will share with you her calm acceptance, and you will be comforted.

Telepathic communication opens up so much joy and pleasure, it more than makes up for the times when you experience pain or sadness with your animal friends. Even then, it is a burden shared and a burden lightened. It enriches life in too many ways to count—all I can say is, please do give it a try.

Just plunge in using these suggestions, or take one of the many workshops and seminars offered by animal communicators everywhere. Join a practice group and take turns talking to each other's companion animals, or start a practice group yourself. However you choose to do it will be right for you.

Let me tell you one last story, to give you a vision of the kind of world it will be when humans and animals talk together freely and openly. It's a vision I had myself after doing a consultation with a sweet young girl named Carlysle and her equally sweet little quarter-horse, Casby.

Carlysle was a member of a pony club. She and Casby did *everything* together—jumping, flat work, gymkana. Intrepid little Casby did everything Carlysle asked of her, but one thing she would not do was allow herself to be shod. Carlysle could ask her to pick up her back feet, and she would, but the shoer couldn't even get near her. Casby would become quite threatening if he even tried. It was a serious problem because they rode in a sand arena and her feet were getting terribly worn down without shoes.

I talked with Casby and learned that early in her training she had encountered some "cowboy" types who had traumatized her. They had tied one of her hind legs up in a very contracted position, leaving her that way, she said, for "hours." It may not have been literally hours, but it was way too long for Casby.

I should say that this kind of training is not done today, or at least I hope it isn't. And there was even worse. After this torture session they had used their ropes to throw her and then sat on her—a completely unnecessary dominance technique. This was all supposed to get her ready to be shod, but obviously it had had exactly the opposite effect.

Now Casby showed me a picture of the shoer who came to her barn. There he was with a cowboy hat, a cowboy belt, cowboy boots—to Casby, he looked just like the feared trainers who had mistreated her. "He is never getting near me!" she said emphatically.

This was the information Carlysle and I needed, and I really don't see how she could have gotten it without telepathic communication. Now that she knew what the problem was, Carlysle learned TEAM bodywork and used it to help Casby release the stored-up trauma and fear in her back legs. At my suggestion she also got a common, household hammer, and whenever she cleaned Casby's back feet she just tapped lightly on her feet with the hammer, to get Casby used to the tapping she'd feel when she was being shod. Then we felt Casby was ready to try again.

I spoke with Casby the night before the shoer was due to come to her barn. I told her that we'd arranged for him to take off his big cowboy hat before he approached her. "Casby," I said, "when he takes off his hat I want you to turn and have a really good look at him. You'll see it's not the same person who hurt you before."

Carlysle told me that that's exactly what happened. The nice shoer removed his hat; Casby turned around, looked deeply into his eyes, and then fetched up a huge sigh. Everyone heard it, and her relief was palpable. She held up her little feet to be shoed, as sweet and dainty as kiss my hand.

It was a beautiful moment, and I know in my heart that it's the way our relationship with animals is meant to be. Whether you want to know what your cat thinks of your dog, or help your horse, or chat with your bird, or commune with a wolf, animal communication is your birthright. It's a blessing just waiting for you. Come, take it.